VOLUME FIFTY TWO

ADVANCES IN CHEMICAL ENGINEERING

Bridging Scales in Modelling and Simulation of Non-Reacting and Reacting Flows. Part I

ADVANCES IN CHEMICAL ENGINEERING

VOLUME FIFTY TWO

ADVANCES IN CHEMICAL ENGINEERING

Bridging Scales in Modelling and Simulation of Non-Reacting and Reacting Flows. Part I

Edited by

ALESSANDRO PARENTE

Combustion and Robust Optimisation (BURN) research group
Aero-Thermo-Mechanics Department
Université Libre de Bruxelles
Bruxelles, Belgium

JURAY De WILDE

Institute of Mechanics, Materials and Civil Engineering (iMMC)
Materials & Process Engineering (IMAP)
Université catholique de Louvain
Louvain-la-Neuve, Belgium

ELSEVIER

ACADEMIC PRESS
An imprint of Elsevier

Academic Press is an imprint of Elsevier
50 Hampshire Street, 5th Floor, Cambridge, MA 02139, United States
525 B Street, Suite 1800, San Diego, CA 92101-4495, United States
The Boulevard, Langford Lane, Kidlington, Oxford OX5 1GB, United Kingdom
125 London Wall, London, EC2Y 5AS, United Kingdom

First edition 2018

Notices
Knowledge and best practice in this field are constantly changing. As new research and experience broaden our understanding, changes in research methods, professional practices, or medical treatment may become necessary.

Practitioners and researchers must always rely on their own experience and knowledge in evaluating and using any information, methods, compounds, or experiments described herein. In using such information or methods they should be mindful of their own safety and the safety of others, including parties for whom they have a professional responsibility.

To the fullest extent of the law, neither the Publisher nor the authors, contributors, or editors, assume any liability for any injury and/or damage to persons or property as a matter of products liability, negligence or otherwise, or from any use or operation of any methods, products, instructions, or ideas contained in the material herein.

ISBN: 978-0-12-815096-2
ISSN: 0065-2377

Publisher: Zoe Kruze
Acquisition Editor: Sam Mahfoudh
Editorial Project Manager: Shellie Bryant
Production Project Manager: Vignesh Tamil
Cover Designer: Greg Harris

Typeset by SPi Global, India

CONTENTS

Contributors *vii*
Preface *ix*

1. Quadrature-Based Moment Methods for Multiphase Chemically Reacting Flows **1**

Rodney O. Fox

1. Introduction 2
2. Quadrature-Based Moment Methods 12
3. Kinetic-Based Finite-Volume Methods 17
4. Application to Turbulent Reacting Flows 22
5. Application to the Population Balance Equation for Fine Particles 30
6. Application to the Kinetic Equation for Gas–Particle Flows 33
7. Concluding Remarks 46

References 46
Further Reading 49

2. Numerical Simulation of Multiphase Reactive Flows **51**

Ziad Hamidouche, Enrica Masi, Pascal Fede, Renaud Ansart, Hervé Neau, Mehrdji Hemati, and Olivier Simonin

1. Introduction 55
2. Brief Description of the Experimental Campaign 57
3. Modeling Description 59
4. Numerical Simulations 65
5. Results and Discussions 75
6. Conclusion 108

Acknowledgments 109
Appendix. Modeling Reactive Gas–Particle Flows 109
References 121

3. Simulation of Turbulent Coalescence and Breakage of Bubbles and Droplets in the Presence of Surfactants, Salts, and Contaminants **125**

Salvatore Falzone, Antonio Buffo, Marco Vanni, and Daniele Luca Marchisio

1. Introduction 126
2. Phenomenology of Fluid Particle Breakup 128
3. Phenomenology of Fluid Particle Coalescence 131

4. Analysis of Different Models for Fluid Particle Breakage Kernel 136
5. Models for the Daughter Size Distribution Function (DaSD) 159
6. Analysis of Different Models for Fluid Particle Coalescence Kernel 163
7. Conclusions 182
References 183

Index *189*

CONTRIBUTORS

Renaud Ansart
Laboratoire de Génie Chimique, LGC, Université de Toulouse, CNRS-Toulouse, France

Antonio Buffo
Institute of Chemical Engineering, Politecnico di Torino, Torino, Italy

Salvatore Falzone
Institute of Chemical Engineering, Politecnico di Torino, Torino, Italy

Pascal Fede
Institut de Mécanique des Fluides de Toulouse, IMFT, Université de Toulouse, CNRS-Toulouse, France

Rodney O. Fox
Iowa State University, Ames, IA, United States

Ziad Hamidouche
Institut de Mécanique des Fluides de Toulouse, IMFT, Université de Toulouse, CNRS-Toulouse, France

Mehrdji Hemati
Laboratoire de Génie Chimique, LGC, Université de Toulouse, CNRS-Toulouse, France

Daniele Luca Marchisio
Institute of Chemical Engineering, Politecnico di Torino, Torino, Italy

Enrica Masi
Institut de Mécanique des Fluides de Toulouse, IMFT, Université de Toulouse, CNRS-Toulouse, France

Hervé Neau
Institut de Mécanique des Fluides de Toulouse, IMFT, Université de Toulouse, CNRS-Toulouse, France

Olivier Simonin
Institut de Mécanique des Fluides de Toulouse, IMFT, Université de Toulouse, CNRS-Toulouse, France

Marco Vanni
Institute of Chemical Engineering, Politecnico di Torino, Torino, Italy

PREFACE

In the context of the International Francqui Chair of Rodney O. Fox, Professor at Iowa State University, two volumes of Advances in Chemical Engineering are dedicated to "Bridging Scales in Modelling and Simulation of Reacting Flows." As Professor Rodney Fox, the contributors to this issue have made numerous ground-breaking contributions to the field of multiphase and reactive flow modeling. The impact of the advances made extends far beyond chemical engineering and touches every technological area dealing with turbulent flow and chemical reactions (e.g., combustion, atmospheric science, and nuclear fuel processing).

All contributions to this collection recognize that future scientific and technological developments will greatly benefit from detailed simulations and that a fundamental approach combining numerical simulations, experiments, and rigorous validation needs to be developed. Scale-bridging approaches are essential for simulations at industrial scale as they allow using the information gathered at the micro scale for more accurate calculations at the macro scale. This is not a trivial task, and implies a thorough understanding of the physical processes involved and their interactions, as well as the possible degree of simplification of the problem, in view of the multidisciplinary application.

Whereas Volume 52 addresses the coupling between flow and species composition fields, Volume 53 focuses on fluidized bed reactors and the advances made in developing scale-bridging approaches. The methods presented in the different chapters are illustrated with industrially relevant applications.

Alessandro Parente and Juray De Wilde

CHAPTER ONE

Quadrature-Based Moment Methods for Multiphase Chemically Reacting Flows

Rodney O. Fox[1]
Iowa State University, Ames, IA, United States
[1]Corresponding author: e-mail address: rofox@iastate.edu

Contents

1. Introduction 2
   1.1 Multiscale Models for Polydisperse Fluid–Particle Flows 3
   1.2 Multiscale Modeling Approach 5
   1.3 Moment Methods 9
2. Quadrature-Based Moment Methods 12
   2.1 Quadrature Method of Moments 13
   2.2 Extended Quadrature Method of Moments 14
   2.3 Hyperbolic Quadrature Method of Moments 15
3. Kinetic-Based Finite-Volume Methods 17
   3.1 Solution of Moment Transport Equations 18
   3.2 Kinetic-Based Spatial Fluxes 19
   3.3 Hyperbolic Spatial Fluxes Using HyQMOM 20
   3.4 Realizable Time-Stepping Schemes 21
4. Application to Turbulent Reacting Flows 22
   4.1 Univariate Case Without Reactions 24
   4.2 Multivariate Case With Chemical Reactions 26
   4.3 Algorithm for Turbulent Reacting Flows 28
5. Application to the Population Balance Equation for Fine Particles 30
   5.1 Fine Particles in Laminar Flows 31
   5.2 Fine Particles in Turbulent Flows 32
6. Application to the Kinetic Equation for Gas–Particle Flows 33
   6.1 Governing Equations for All Gas–Particle Flow Regimes 35
   6.2 Solution Algorithm for All Gas–Particle Flow Regimes 39
   6.3 Application to Particle-Laden Flows 42
7. Concluding Remarks 46

References 46

Further Reading 49

Advances in Chemical Engineering, Volume 52
ISSN 0065-2377
https://doi.org/10.1016/bs.ache.2018.01.001

Abstract

The numerical simulation of multiphase chemically reacting flows is very challenging due to their multiscale nature. In this chapter, the focus is on disperse multiphase flows with a continuous phase (gas or liquid) surrounding one or more disperse phases (e.g., particle, drops, or bubbles). In such flows, the disperse phase is usually characterized by a size, concentration, and/or temperature distribution that must be accounted for in the mathematical model. In this context, a population balance model (PBM) for the number density function is the most convenient way to account for polydispersity. In addition, the continuous phase is often turbulent, so mixing and chemical reactions in a turbulent flow must also be modeled. This can be done using probability density function (PDF) methods. In this chapter, computational methods for approximating solutions to PBM and PDF transport equations are reviewed in the context of computational fluid dynamic (CFD). In particular, quadrature-based moment methods (QBMM) have proven to be accurate and numerically efficient when combined with CFD simulations. Applications with increasing degrees of complexity in the flow physics are introduced to illustrate how QBMM are used to solve real problems in chemical engineering.

1. INTRODUCTION

The overall objective of this chapter is to provide the reader with a (nearly) self-contained overview of the current status of quadrature-based moment methods (QBMM) as applied to multiphase and/or chemically reacting flows. The first such application dates to McGraw (1997), and in the past 20 years considerable progress has been made in a number of different directions. The first applications of QBMM were to "fine particles" (e.g., aerosols), which are flow followers, in the context of (turbulent) transport. These applications required QBMM to close the source terms for aggregation, breakage, etc. (Marchisio et al., 2003a,b,c; Nguyen et al., 2017; Wick et al., 2017; Zucca et al., 2006), or for micromixing and chemical reactions (Fox, 2003; Madadi-Kandjani et al., 2017; Piton et al., 2000; Wang and Fox, 2003a,b). The extension to inertial particles with finite Stokes number occurred later (Fan et al., 2004; Marchisio and Fox, 2005), but has taken longer to develop because of the particular nature of the spatial fluxes in kinetic equations (Chalons et al., 2010b, 2017; Fox et al., 2017). Likewise, a robust extension of QBMM to multivariate distribution functions required much work (Cheng and Fox, 2010; Cheng et al., 2010; Fox, 2006, 2009a; Yuan and Fox, 2011) (and can undoubtedly still be improved), as did the extension to continuous distributions (Chalons et al., 2010a; Yuan et al., 2012). Along the way, it became evident for spatially inhomogeneous flows that improvements in finite-volume schemes to make them "realizable" when solving moment sets could be achieved with QBMM (Laurent and Nguyen, 2017; Vikas et al., 2011a), making Eulerian methods competitive against Lagrangian methods for computational fluid dynamics (CFD) (de Chaisemartin et al., 2008, 2009; Fox et al., 2008; Kah et al., 2010). In the following sections, many of these points will be reviewed, especially those that have advanced significantly beyond the material provided in Marchisio and Fox (2013).

It is now generally acknowledged that, given the difficulty of obtaining experimental data for multiphase flows, detailed numerical simulations are a promising tool for understanding the impact of operating conditions and physical properties on the flow dynamics. However, even the "simplest" case with monodisperse particles of known size and shape in a continuous fluid phase is challenging to simulate across the full range of scales and parameters of interest (Marchisio and Fox, 2013). In this chapter, a brief overview of the challenges and modeling approaches for dealing with fluid–particle flows is presented. For completeness, we include the possibility that the fluid phase is turbulent, so that a multiscale turbulence model is needed to account for the effect of the unresolved scales on the resolved scales (e.g., Reynolds-averaged quantities or filtered quantities in large-eddy simulations). However, for simplicity, here we show and discuss only the model equations for homogeneous chemical reactions in the fluid phase.

Finally, the reader should keep in mind that QBMM is a toolbox for *solving* transport equations involving distribution functions (e.g., probability density function or number density functions). Thus, the correct physics and chemistry needed to treat a particular application must already be embedded in the transport equation (see Marchisio and Fox, 2013, for specific examples). In other words, QBMM will not provide you with a "better" description of your physical system, but instead will provide you with a tool to approximate solutions for a distribution function involving many independent variables. As such, some expertise is required to choose the "best" tools from the QBMM toolbox for a given problem. By using specific examples, in the following sections we attempt to illustrate how this is done.

1.1 Multiscale Models for Polydisperse Fluid–Particle Flows

The numerical simulation of polydisperse fluid–particle flows arising in technological and environmental applications is challenging due to their multiscale nature (see Fig. 1 for an example). In this chapter, we will use

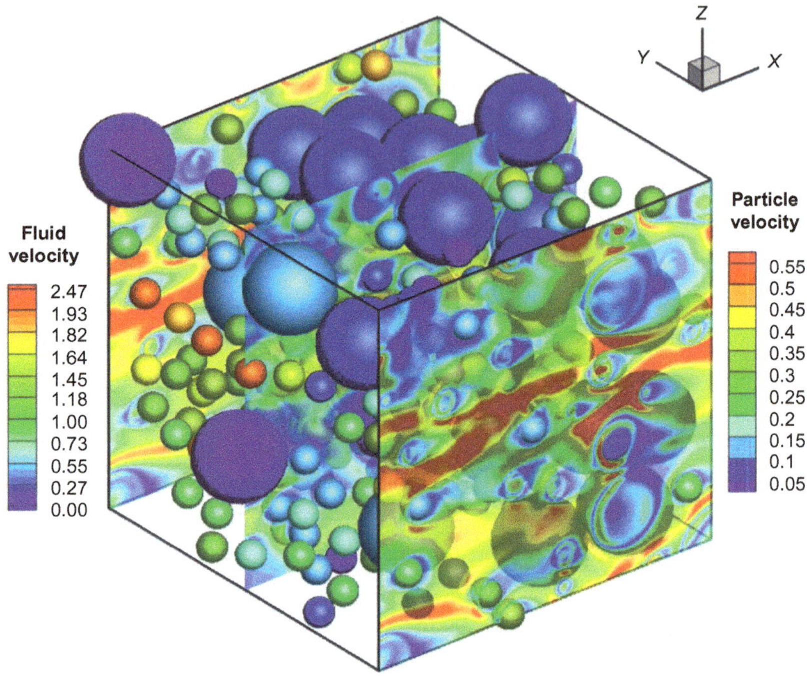

Fig. 1 Bidisperse gas–particle flow (PR-DNS of Subramaniam group). Large (small) inertial particles have different mean and fluctuating velocities, and the fluid velocity varies significantly inside the computational domain due to momentum coupling. The variance of the particle velocity about the cell-mean velocity is measured by the granular temperature, while for the fluid phase it is referred to as "pseudo-turbulent" kinetic energy. Here, "pseudo-turbulence" refers to fluid velocity fluctuations at the scale of the particles. On even larger scales, multiphase turbulence can be generated by the momentum and volume fraction coupling between the two phases (Fox, 2014).

the term "particle" to refer to a disperse phase (in-miscible liquid or solid) present in a continuous fluid phase. Thus, the disperse phase might be a solid particle, liquid droplet, or a gas bubble. The principal characteristics of disperse multiphase flows are the following:

(a) continuous fluid phase (Newtonian, non-Newtonian, etc.),
(b) chemical reactions in continuous fluid phase modified by turbulence,
(c) disperse particle phase (solid particle, droplet, bubbles, etc.),
(d) size (chemical composition, etc.) distribution of particles,
(e) nonnegligible particle inertia (as measured by a nonzero Stokes number),
(f) inelastic collisions (particle–particle, particle–wall),
(g) variable mass loading (e.g., very dilute risers to dense fluidized beds),

(h) multiphase turbulence (due to mean shear and momentum and/or volume fraction coupling between phases).

In general, we distinguish between noninertial and inertial particles. The former are typically submicron (or nanoscale) and essentially follow the fluid flow. The later are larger, so we must account for their momentum and its coupling with the fluid phase. In either case, particle–particle interactions can result in aggregation and breakage, so at a minimum we must model the particle size distribution (PSD). Moreover, the particles may have different chemical compositions, or temperatures, so, in addition to size, other so-called internal variables are needed to describe the particle phase.

The complex physics and chemistry of disperse multiphase flows results in many modeling challenges. The most significant among these are the following.

(a) Strong coupling between the fluid and particle phases.
(b) Wide range of particle volume fractions (even in same flow).
(c) Inertial particles with a wide range of Stokes numbers.
(d) Collision-dominated to collision-less regimes in same flow.
(e) Granular temperature of disperse phase can be small and large in same flow.
(f) Polydispersity (e.g., size, density, shape) is almost always present.

In order to deal with all of these challenges, we need a multiscale modeling framework that can handle all aspects.

1.2 Multiscale Modeling Approach

The kinetic-theory modeling approach (Perthame, 2003) for the disperse phase is summarized in Fig. 2. The principal feature is the formulation of a mesoscale model with physics-based closures derived from the "exact" microscale model. For example, particle–particle collisions are resolved at the microscale, but must be modeled at the mesoscale, because information about how the two particles' velocities are correlated before collision is missing (Marchisio and Fox, 2013). The mesoscale model can have a large number of phase-space variables to describe the particle size distribution, particle velocity distribution, etc.

A tractable macroscale model for large-scale simulations is sought by using mathematical closures of the mesoscale model. A well-known example of this procedure is the derivation of the Navier–Stokes equation in the kinetic theory of ideal gases where the closed mesoscale model is the Boltzmann equation. For disperse multiphase flows, the kinetic-based

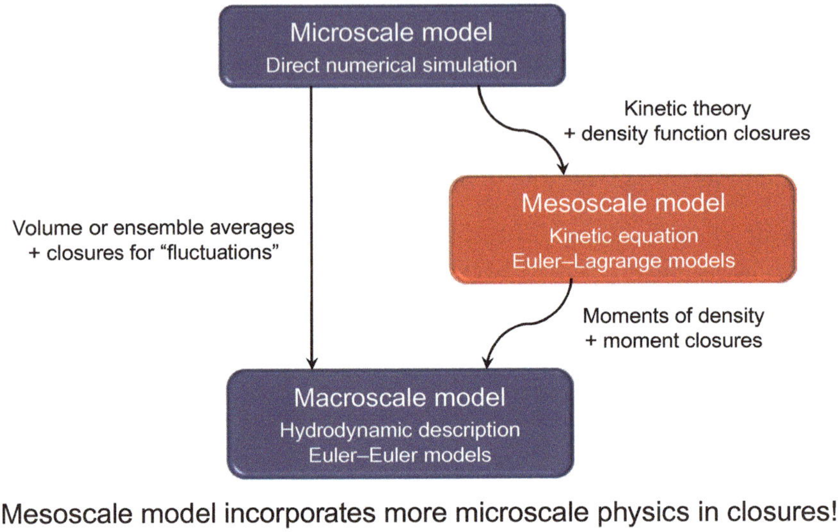

Fig. 2 The kinetic-theory modeling approach uses microscale DNS (see Fig. 1) to develop a closed mesoscale model that captures the relevant physics of the disperse phase. Macroscale computational fluid dynamics (CFD) models are derived from the mesoscale model by using mathematical closures in terms of the moments. If the fluid is turbulent, an analogous multiscale model is developed using PDF methods. The mesoscale model for the fluid is then a PDF transport equation for the fluid composition vector. In either case, the mesoscale model has high dimensionality due to the "internal" or "phase-space" variables. In order to make the solution tractable in the context of CFD, this dimensionality must be reduced, which can be accomplished by solving for a small set of moments of the distribution function.

model is applied to the particle phase. If the fluid phase is turbulent, a probability density function (PDF) approach is used to model its statistical properties. The PDF and kinetic model are coupled because of mass, momentum, and energy exchange between the two phases. In each case, a multiscale model that accounts for the effect of unresolved scales on the resolved scales is required. Here, we will refer to such models as *mesoscale*.

In fluid–particle flows, the mesoscale model equations for the fluid and particle phases typically has one of the four following forms (repeated Roman indices imply summation over all components):

1. *Composition PDF transport equation* (fluid phase) for the PDF $f(t, \mathbf{x}, \boldsymbol{\psi})$ where $\boldsymbol{\phi}$ is fluid composition vector (chemical species, temperature) and $\boldsymbol{\psi}$ is the corresponding phase-space vector:

$$\frac{\partial f}{\partial t} + \frac{\partial \langle u_i | \boldsymbol{\phi} = \boldsymbol{\psi} \rangle f}{\partial x_i} + \frac{\partial}{\partial \psi_i}[S_i(\boldsymbol{\psi}) f] = \mathbb{M}$$

with closed chemical reactions $\mathbf{S}$ and a micromixing model $\mathbb{M}$. In words, the PDF is the probability of finding fluid particles with composition $\boldsymbol{\psi}$ at spatial location $\mathbf{x}$ at time t. The conditional velocity $\langle \mathbf{u} | \boldsymbol{\phi} = \boldsymbol{\psi} \rangle$ is usually modeled by a gradient-diffusion model:

$$\langle u_i | \boldsymbol{\phi} = \boldsymbol{\psi} \rangle = \langle u_i \rangle - \frac{D_t}{f} \frac{\partial f}{\partial x_i}$$

where D_t is the turbulent diffusivity. The simplest micromixing model is the interaction-by-exchange-with-the-mean (IEM) model

$$\mathbb{M} = \frac{\partial}{\partial \psi_i} \left[\frac{1}{\tau_\phi} (\psi_i - \langle \phi_i \rangle) f \right]$$

where τ_ϕ is the micromixing timescale. The definitions of both D_t and τ_ϕ depend on the turbulence model used to describe the unresolved scales in the fluid phase. In a multiphase flow, the composition PDF transport equation contains an additional term to describe the changes in the PDF due to mass transfer between the two phases.

2. *Population balance equation* (PBE for fine particles) for the number density function (NDF) $n(t, \mathbf{x}, \boldsymbol{\xi})$ where $\boldsymbol{\xi}$ is, for example, particle size or composition:

$$\frac{\partial n}{\partial t} + \frac{\partial u_i n}{\partial x_i} + \frac{\partial}{\partial \xi_i} [G_i(t, \mathbf{x}, \mathbf{v}, \boldsymbol{\xi}) \, n] = \mathbb{B} + \mathbb{J}$$

with growth $\mathbf{G}$, aggregation/breakage $\mathbb{B}$, and nucleation/dissolution operator $\mathbb{J}$. In words, the NDF is the number concentration of particles with internal variables $\boldsymbol{\xi}$ at spatial location $\mathbf{x}$ at time t. Here, the growth and nucleation terms are coupled to the fluid phase due to mass transfer. For turbulent flows, $\mathbf{u}$ is replaced by a turbulence model for the fluid velocity, e.g., a gradient-diffusion model

$$u_i \approx \langle u_i \rangle - \frac{D_t}{n} \frac{\partial n}{\partial x_i}.$$

In this case, n is the Reynolds-averaged NDF, and the models appearing in $\mathbf{G}$, $\mathbb{B}$, and $\mathbb{J}$ must be formulated in terms of the resolved fluid-phase turbulent statistics. In some cases, it is necessary to account for fluctuations in n about the Reynolds-average, and to include a micromixing model. For example, when nucleation and aggregation are very fast relative to turbulent mixing, such fluctuations can have a decisive effect on the shape of the PSD (Wang and Fox, 2003a).

3. *Kinetic equation* (KE for monodisperse inertial particles) for the NDF $n(t, \mathbf{x}, \mathbf{v})$ where $\mathbf{v}$ is particle velocity:

$$\frac{\partial n}{\partial t} + \frac{\partial v_i n}{\partial x_i} + \frac{\partial}{\partial v_i}[A_i(t, \mathbf{x}, \mathbf{v})\, n] = \mathbb{C}$$

with closed acceleration $\mathbf{A}$ and particle–particle collision operator $\mathbb{C}$ (e.g., Boltzmann equation). In words, the NDF is the number concentration of particles with velocity $\mathbf{v}$ at spatial location $\mathbf{x}$ at time t. Here, the acceleration term is coupled to the fluid phase due to drag. If the fluid phase is turbulent, then the model for $\mathbf{A}$ must account for the resolved and unresolved fluid velocity components. The mesoscale model for collisions may need to account for finite-sized particles (e.g., Boltzmann–Enskog equation) and energy dissipation. In granular flows, the frequency of collisions is proportional to the square-root of the granular temperature multiplied by the particle-phase volume fraction. When collisions are infrequent, the shape of the NDF can be far from Gaussian. In such cases, the NDF is said to be far from equilibrium.

4. *Generalized population balance equation* (GPBE for polydisperse inertial particles) for the NDF $n(t, \mathbf{x}, \mathbf{v}, \boldsymbol{\xi})$ where $\boldsymbol{\xi}$ is, for example, particle mass or particle composition:

$$\frac{\partial n}{\partial t} + \frac{\partial v_i n}{\partial x_i} + \frac{\partial}{\partial v_i}[A_i(t, \mathbf{x}, \mathbf{v}, \boldsymbol{\xi})\, n] + \frac{\partial}{\partial \xi_i}[G_i(t, \mathbf{x}, \mathbf{v}, \boldsymbol{\xi})\, n] = \mathbb{C} + \mathbb{J}$$

with closed acceleration $\mathbf{A}$, "growth" $\mathbf{G}$, collision/aggregation/breakage $\mathbb{C}$, and nucleation/evaporation operator $\mathbb{J}$. In words, the NDF is the number concentration of particles with velocity $\mathbf{v}$ and internal variables $\boldsymbol{\xi}$ at spatial location $\mathbf{x}$ at time t. Here, in addition to the terms contained in the KE, the "growth" term $\mathbf{G}$ is coupled to the continuous fluid phase due to mass transfer. Solving the GPBE is particularly challenging due to the large number of independent variables.

The equations for the distribution functions given above are coupled to the Navier–Stokes equation for the fluid phase. The multiscale models appearing in the PBE, KE, and GPBE are derived to be consistent with the microscale model for the particles and, hence, the functional dependencies of the acceleration, growth, collision/aggregation operators are *known*, but often very nonlinear. In other words, the equations for the distribution functions are well-posed with known initial and boundary conditions that are derived from the microscale physics. Therefore, in principle, the mesoscale model can be solved exactly for the PDF or NDF. However, in practice, even a numerical solution can be intractable due to the large number of independent variables and the complexity of the flow. For large-scale simulations, a simpler approach is needed, preferably one that can be coupled with an existing CFD code.

1.3 Moment Methods

One powerful approach for developing macroscale models (which is the focus of this chapter) is referred to as *moment methods*. The starting point is the Eulerian moment transport equation, which can be derived for the mesoscale equations in Section 1.2 (see Marchisio and Fox, 2013, for details). This procedure is described next.

1. *Composition moment transport equation* (univariate) for the k-order moment $\langle \phi^k \rangle := \int \psi^k f \, d\psi$ is given by

$$\frac{\partial \langle \phi^k \rangle}{\partial t} + \frac{\partial \langle u_i \phi^k \rangle}{\partial x_i} - k \langle S(\phi) \rangle = \int \psi^k \mathbb{M} \, d\psi$$

   where the convective flux $\langle u_i \phi^k \rangle$ and chemical source term $\langle S(\phi) \rangle$ usually require closure. If a gradient-diffusion model is used for the flux, then

$$\langle u_i \phi^k \rangle = \langle u_i \rangle \langle \phi^k \rangle - D_t \frac{\partial \langle \phi^k \rangle}{\partial x_i}$$

   is closed. Likewise, with the IEM model, the micromixing term is closed:

$$\int \psi^k \mathbb{M} \, d\psi = -\frac{k}{\tau_\phi} \left(\langle \phi^k \rangle - \langle \phi \rangle \langle \phi^{k-1} \rangle \right).$$

   The principal challenge is thus to close the chemical source term given knowledge of a finite set of moments. In most applications, a multivariate composition vector is required, which significantly increases the number of moments needed for closure.
2. *KE*: In 1-D, the velocity moments are $M_k := \int v^k n \, dv$. Integration over the KE yields the moment transport equation:

$$\frac{\partial M_k}{\partial t} + \frac{\partial M_{k+1}}{\partial x} = k \int v^{k-1} A n \, dv + \int v^k \mathbb{C} \, dv$$

   where the first three moments are[a] $M_0 = \rho_p \alpha_p$, $M_1 = \rho_p \alpha_p U_p$, and $M_2 = \rho_p \alpha_p (U_p^2 + \Theta_p)$. Here, α_p is the particle-phase volume fraction, U_p is the

[a] For monovariate particles, the particle mass times the number density is M_0. Therefore, it is convenient to define the NDF in terms of the mass density function.

particle-phase velocity, and Θ_p is the granular temperature. The flux M_{k+1} is referred to as the free-transport (or kinetic) flux of M_k and is unclosed for the largest k. The terms on the right-hand side are source terms due to acceleration and particle–particle collisions. In some cases, the right-hand-side terms may be closed. For example, with Stokes drag:

$$\int v^{k-1} A n \, \mathrm{d}v = \frac{1}{\tau_p}(u_f M_{k-1} - M_k)$$

where u_f is the fluid velocity and τ_p is the drag time scale. Likewise, if a linear collision model such a BGK is used, then the collision term will depend on known moments. Thus, the most challenging term to close is usually the kinetic flux. In applications, the fluid velocity $\mathbf{u}_f$ is 3-D and must be found by solving a Navier–Stokes equation, which is nontrivial.

3. *GPBE*: In 1-D, the joint velocity–mass moments are defined by $M_{k,l} := \int v^k \xi^l n \, \mathrm{d}v\mathrm{d}\xi$. Integrating over the GPBE yields the moment transport equation:

$$\frac{\partial M_{k,l}}{\partial t} + \frac{\partial M_{k+1,l}}{\partial x} = k \int v^{k-1} \xi^l A n \, \mathrm{d}v\mathrm{d}\xi + l \int v^k \xi^{l-1} G n \, \mathrm{d}v\mathrm{d}\xi + \int v^k \xi^l \mathbb{C} \, \mathrm{d}v\mathrm{d}\xi$$

where some of the lower-order moments are $M_{0,1} = \rho_p \alpha_p$, $M_{1,1} = \rho_p \alpha_p U_p$, and $M_{2,1} = \rho_p \alpha_p (U_p^2 + \Theta_p)$. Moments $M_{0,0}$, $M_{0,1}$, $M_{0,2}$, etc. are needed for the mass distribution, while moments with $k > 0$ are needed to account for the correlation between the particle mass and velocity for finite-Stokes-number particles. Because the drag time scale $\tau_p(\xi)$ depends on the particle size, even with Stokes drag the acceleration term will not be closed in terms of the moments.

In general, the terms above in red require mathematical closure because the distribution function is unknown. However, if we can approximate the PDF/NDF from knowledge of a finite set of its moments, then these equations will be closed. The classical example is the *hydrodynamic model* where we assume that the velocity NDF in the KE is a functional of α_p, U_p, and Θ_p, which leads to the gas dynamics equations for mass, momentum, and energy. The hydrodynamic model will be valid when the velocity NDF is mainly determined by the collision operator. In the absence of collisions (e.g., dilute flows or negligible granular temperature), a more general moment-closure method is needed. (See Fig. 3 for an example of particle trajectory crossing without collisions.) Closure with QBMM is shown schematically in Fig. 4 and is the main focus of this chapter.

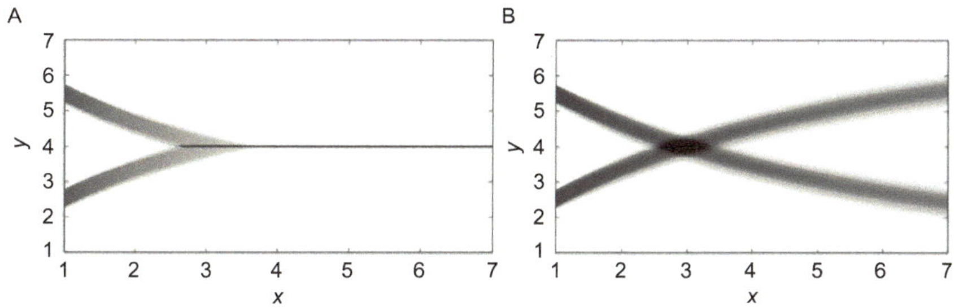

Fig. 3 Example of a nonequilibrium solution to the collision-less KE. Particles are injected into the domain at $x = 1$ with zero velocity dispersion. (A) Solution of moment transport equations using moments $\mathbf{M} = \{M_0, M_1\}$. (B) Solution using moments $\mathbf{M} = \{M_0, M_1, M_2, M_3\}$. The trajectory crossing on the right is the exact solution. At the crossing point, the velocity NDF is bimodal (i.e., sum of two weighted delta functions). Quadrature-based moment methods with only one quadrature node (A) produce a delta-shock when the jets try to cross. However, with two nodes (B) the exact solution can be captured. In general, to capture the crossing of N jets, at least $2N$ velocity moments (i.e., $\mathbf{M} = \{M_0, M_1, \ldots, M_{2N-1}\}$) must be included in the moment system. If collisions are present, less moments are required. In the case of highly collisional flow, only moments $\mathbf{M} = \{M_0, M_1, M_2\}$ are needed (i.e., the hydrodynamic model).

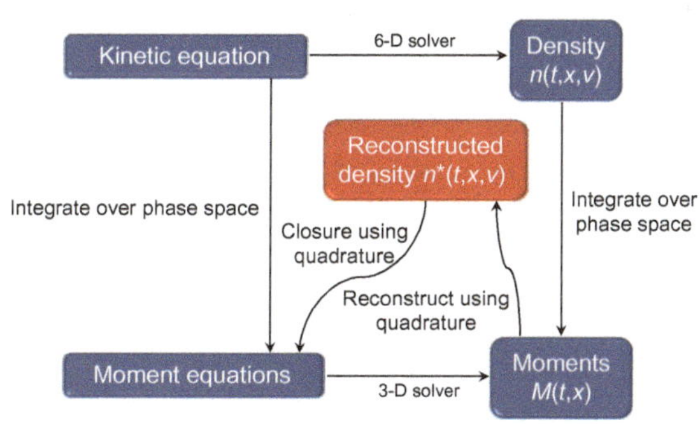

Close moment equations by reconstructing density function

Fig. 4 Quadrature-based moment methods reconstruct the PDF/NDF from a finite sets of its moments. The reconstruction takes the form of a sum of weighted delta functions that exactly reproduces the known moments in each grid cell at each time step. Here, the 3-D solver can be a classical finite-volume CFD code. The 6-D solver is a direct solver for the PDF/NDF in 3-D physical and 3-D phase space. In most applications, direct knowledge of the form of the NDF is not needed as we are mainly interested in its lower-order moments (e.g., to compare with experiments). Thus, if the quadrature-based closure is sufficiently accurate, then employing the 3-D solver can be considerably faster. Note, however, that the computational cost of the reconstruction must be low for the method to be successful for realistic applications.

2. QUADRATURE-BASED MOMENT METHODS

The Basic Idea: Given a finite set of moments, reconstruct the NDF. While the basic idea is simple, there are several important things to consider in practice:

(i) Which set of moments should we choose?
(ii) What computational approach should we use for reconstruction?
(iii) How can we extend the approach to multivariate phase space?
(iv) How should we design the numerical solver for the moment transport equations?

In all cases, we must be able to demonstrate a priori that the numerical algorithm is robust and accurate. Another key point is that the method must be very fast because we will need to do the moment inversion in every grid cell at every time step during the CFD simulation. This consideration rules out numerical algorithms based on iterations to converge to the final result (e.g., entropy maximization).

One classical method for approximating 1-D integrals is Gaussian quadrature, which has the following mathematical properties (Gautschi, 2004):

(a) The formula

$$\int g(v)n(v)\,\mathrm{d}v = \sum_{\alpha=1}^{N} n_\alpha g(v_\alpha) + R_N(g)$$

is a Gauss quadrature iff the N nodes v_α are the roots of an Nth-order orthogonal polynomial $P_N(v)$ ($\perp$ with respect to $n(v)$). This quadrature formula is exact (i.e., $R_N(g) = 0$) if $g(v)$ is a polynomial of order less than $2N$.

(b) A robust inversion algorithm for integer moments $M_k = \int v^k n(v)\,\mathrm{d}v$ exists to find the weights and abscissas:

$$\{M_0, M_1, \ldots, M_{2N-1}\} \overset{\text{QMOM}}{\Longrightarrow} \{n_1, n_2, \ldots, n_N\}, \{v_1, v_2, \ldots, v_N\}$$

which we refer to as the quadrature method of moments (QMOM). The algorithm solves a small (N) eigenvalue/eigenvector problem, which is relatively fast and does not require iterations. However, for large N, the algorithm is ill-conditioned (because the difference between M_k and M_{k+1} is very small for large k).

(c) If the moments correspond to a well-defined NDF (i.e., nonnegative), the weights $\{n_1, n_2, \ldots, n_N\}$ will be nonnegative, and the abscissas $\{v_1, v_2, \ldots, v_N\}$ are physically well defined (i.e., they are in the support of the NDF).

The moments of a well-defined 1-D NDF are said to be *realizable*. Using Hankel matrices, it is possible to check if a set of moments is realizable without knowledge of the NDF (Dette and Studden, 1997). From a practical perspective, the numerical solution to the moment transport equation must yield realizable moments in order for us to apply QMOM to find the quadrature weights and abscissas. In general, not all finite-volume methods yield realizable moments so we must develop new numerical methods to ensure realizability. We will refer to these as realizable finite-volume methods, or realizable numerical methods, in general.

2.1 Quadrature Method of Moments

The favorable properties of Gaussian quadrature described above first led to applying QMOM to the size moments of (noninertial) aerosols (McGraw, 1997). In general, the unclosed source terms in the moment equations are approximated as

$$\frac{\mathrm{d}\mathbf{M}}{\mathrm{d}t} = \int_0^{\infty} \mathbf{S}(\xi)n(\xi)\mathrm{d}\xi \approx \sum_{\alpha=1}^{N} n_\alpha \mathbf{S}(\xi_\alpha)$$

where $\mathbf{M} = \{M_0, M_1, \ldots, M_{2N-1}\}$ and $\mathbf{S}$ is a generic "source term." Experience has shown that QMOM applied to source terms has the following properties:

(a) The QMOM closure is exact if $\mathbf{S}$ is polynomial of order $\leq 2N$, otherwise it is a very good approximation.

(b) QMOM provides good approximations with small $N \lessapprox 4$ (e.g., for source terms corresponding to aggregation and breakage).

(c) When solving the moment equations numerically, the moments $\mathbf{M}$ must remain *realizable* for moment inversion, which is almost always the case when the NDF has nonzero variance. Otherwise, care must be taken to ensure that the numerical algorithm does not produce nonrealizable moments.

Note that QMOM is equivalent to reconstructing the NDF as a sum of Dirac delta functions (Dette and Studden, 1997):

$$n^*(\xi) = \sum_{\alpha=1}^{N} n_\alpha \delta(\xi - \xi_\alpha),$$

which is *realizable* because $n_\alpha \geq 0$ for all α. The abscissas ξ_α can be interpreted as weighted samples (or "parcels") that adapt dynamically with the evolution of the moments. In general, the QMOM reconstruction using $\mathbf{M}$ leads to the smallest possible value for M_{2N} (i.e., all other definitions of $n^*(\xi)$ with moments $\mathbf{M}$ will have a larger M_{2N}). From a practical point of view, having a reconstruction in the form of Dirac delta function is particularly convenient for evaluating the integrals appearing in the unclosed terms of the moment transport equation introduced in Section 1.3.

2.2 Extended Quadrature Method of Moments

In practice, it is sometimes desirable to obtain a continuous representation for $n^*(\xi)$ instead of Dirac delta functions. At the same time, we would like to retain the noniterative nature of QMOM as much as possible. For this purpose, we can add an additional moment $\mathbf{M} = \{M_0, M_1, \ldots, M_{2N}\}$ where M_{2N} is used to parameterize the width σ of a kernal density function (KDF), denoted by $\delta_\sigma(\xi - \xi_\alpha)$. For example, the KDF can be a Gaussian PDF centered at ξ_α with standard deviation σ (Chalons et al., 2010a). This extended QMOM (or EQMOM) defines the reconstructed NDF as

$$n^*(\xi) = \sum_{\alpha=1}^{N} n_\alpha \delta_\sigma(\xi - \xi_\alpha).$$

The EQMOM algorithm is based on the QMOM algorithm, but a 1-D nonlinear solver is used to find σ such that M_{2N} is a moment of n^*. If this is not possible, the σ is chosen such that M_{2N} found from n^* is as close as possible to the known value (Nguyen et al., 2017).

In practice, introducing the KDF in EQMOM complicates the evaluation of source terms in the moment transport equations (as compared to Dirac delta functions). For example, the double integration needed for aggregation source terms is much more expensive to evaluate with EQMOM than with QMOM. On the other hand, if EQMOM is needed to obtain a continuous NDF (e.g., for plotting output), then we must solve for all moments $\mathbf{M} = \{M_0, M_1, \ldots, M_{2N}\}$. In order to use all of these moments in a Dirac delta function representation, it is convenient to used Gauss–Radau quadrature wherein one abscissas ξ_0 is fixed (e.g., the lower/upper bound) but its weight n_0 is free. This leads to

$$n^*(\xi) = \sum_{\alpha=0}^{N} n_\alpha \delta(\xi - \xi_\alpha),$$

and a slight variation of the eigenvalue problem solved in QMOM yields $N + 1$ weights and N abscissas (Gautschi, 2004). Thus, inside the simulation code, Gauss–Radau quadrature is used for reconstructing the NDF (and evaluating integrals); while EQMOM is reserved for producing a continuous distribution for output (or when otherwise required).

2.3 Hyperbolic Quadrature Method of Moments

While the Dirac delta function reconstruction works well for source terms, it is not as well suited for closing the spatial flux in the KE and GPBE. Technically, the KE for the velocity NDF is a hyperbolic equation (i.e., the spatial flux has real, distinct eigenvalues), and thus so should be the flux in the moment equations. In practice, this means with given the velocity moments $\mathbf{M} = \{M_0, M_1, \ldots, M_{2N}\}$, we must choose M_{2N+1} such that the moment system

$$\begin{aligned}
\frac{\partial M_0}{\partial t} + \frac{\partial M_1}{\partial x} &= 0 \\
\frac{\partial M_1}{\partial t} + \frac{\partial M_2}{\partial x} &= 0 \\
&\vdots \\
\frac{\partial M_{2N-1}}{\partial t} + \frac{\partial M_{2N}}{\partial x} &= 0 \\
\frac{\partial M_{2N}}{\partial t} + \frac{\partial M_{2N+1}}{\partial x} &= 0
\end{aligned}$$

is hyperbolic (Fox et al., 2017). If QMOM is used to close M_{2N}, then the system will be only weakly hyperbolic (Chalons et al., 2010b), meaning that there will be N distinct eigenvalues corresponding to the abscissas and N eigenvectors (instead of $2N$). The numerical solution to a weakly hyperbolic system can exhibit so-called delta-shocks, which are points $\mathbf{x}$ where $M_0(t, \mathbf{x})$ is infinite. This type of behavior is nonphysical.

In order to avoid delta-shocks, a closure of the moment M_{2N+1} using hyperbolic QMOM is preferable. For velocity moments, the odd-order moment M_{2N+1} can take any value from $\pm\infty$. In practice, it is preferable to fix the central moment C_{2N+1} (i.e., the velocity NDF is recentered at zero mean velocity), which is entirely equivalent to fixing M_{2N+1}. At present (Fox et al., 2017), unique choices for C_{2N+1} are known for $N = 1, 2, 3$, which make the moment system hyperbolic for all realizable moments $\mathbf{M} = \{M_0, M_1, \ldots, M_{2N}\}$. We refer to this closure for the 1-D velocity moments as hyperbolic QMOM or HyQMOM.

In principle, the same ideas can be used for 2-D and 3-D velocity moments. As an example, consider the 2-D KE with velocity moments

$$M_{i,j} = \int u^i v^j n(u,v)\, \mathrm{d}u\mathrm{d}v.$$

In order to handle the second direction in phase space, a variation of the conditional quadrature methods of moments (CQMOM) is adapted to the HyQMOM, leading to CHyQMOM (Fox et al., 2017). For $N = 2$, the reconstructed NDF has the following properties[b] :

(a) CHyQMOM reconstructs the 2-D velocity NDF as

$$n^*(u,v) = M_{0,0} \sum_{\alpha=1}^{3} \sum_{\beta=1}^{3} p_\alpha p_{\alpha\beta}\, \delta(u - \bar{u} - u_\alpha)\delta(v - \bar{v} - \bar{v}_\alpha - v_{\alpha\beta})$$

where $\bar{u} = M_{1,0}/M_{0,0}$, $\bar{v} = M_{0,1}/M_{0,0}$ are the mean velocities. The abscissas u_α, $v_{\alpha\beta}$, and the conditional moment $\bar{v}_\alpha$ are found from the bivariate central moments defined by

$$C_{i,j} = \frac{1}{M_{0,0}} \int_{\mathbb{R}^2} (u - \bar{u})^i (v - \bar{v})^j n(u,v)\, \mathrm{d}u\mathrm{d}v.$$

The probabilities p_α are found using 1-D HyQMOM, while the $p_{\alpha\beta}$ are found from 2-D CQMOM (Yuan and Fox, 2011).

(b) In 2-D velocity phase space, the nine quadrature nodes are found from the 10 moments:

$$\mathbf{M} = \begin{pmatrix} M_{0,0} & M_{0,1} & M_{0,2} & M_{0,3} & M_{0,4} \\ M_{1,0} & M_{1,1} & & & \\ M_{2,0} & & & & \\ M_{3,0} & & & & \\ M_{4,0} & & & & \end{pmatrix}.$$

A 9-node, higher-order reconstruction using all of the third-order moments is also possible (by changing the definition of $\bar{v}_\alpha$) (Fox et al., 2017).

Like CQMOM, the CHyQMOM reconstruction can reproduce particle trajectory crossings for $1 < N$, but is hyperbolic (i.e., real and distinct eigenvalues) for the spatial fluxes of $\mathbf{M}$. Thus, solutions to the moment system do not exhibit delta-shocks nor other nonphysical behavior. In practice, operator splitting is used in the simulation code to treat the spatial fluxes separately from the source terms. In other words, for the 1-D moment set

[b] Note that for the definition of N used in this section, the number of HyQMOM abscissas is $N + 1$. Thus, $N = 2$ corresponds to 3 abscissas. In the original paper (Fox et al., 2017), N is used to denote the number of abscissas and the transported moment set is $\mathbf{M} = \{M_0, M_1, \ldots, M_{2N-2}\}$.

$\mathbf{M} = \{M_0, M_1, ..., M_{2N}\}$, HyQMOM is first used to update the moments due to the spatial fluxes, and then Gauss–Radau quadrature is used to update the moments due to the source terms. In both steps, care must be taken to ensure that the numerical algorithm is realizable.

3. KINETIC-BASED FINITE-VOLUME METHODS

Because classical high-order finite-volume (FV) schemes do not ensure realizability, kinetic-based FV (KBFV) methods have been developed to ensure that in CFD codes the numerical solution for the moments are realizable (Marchisio and Fox, 2013). The key issue is centered on how the moment vector is represented at the interface between two finite volumes (see Fig. 5 for details). The simplest (but inaccurate) approach is to assume that the moments are constant in each volume (with a jump at the interface). This is the so-called first-order upwind method that is well known to produce too much artificial "diffusion." In general, a second-order method is preferred, but all classical second-order methods will yield unrealizable moments due to discretization errors (Laurent and Nguyen, 2017; Vikas et al., 2011a, 2013b). Here, we describe the basic problem using a 2-D moment set.

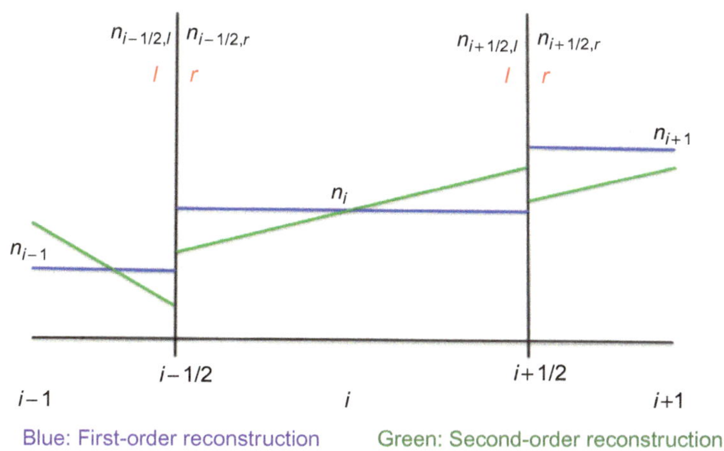

Fig. 5 Neighboring finite-volume cells in a 1-D domain. The NDF in cell i is denoted by $n_i(x)$. The interfaces between cells are at $i \pm \frac{1}{2}$ and the *left* (l) and *right* (r) sides of the interfaces are shown. Inside each cell, $n_i(x)$ must be approximated from the known values of the cell-average moments $\mathbf{M}_i$. The first-order approximation takes n_i as constant. The second-order approximation takes n_i as linear in x.

3.1 Solution of Moment Transport Equations

Given the transported moments **M**, we wish to solve the 2-D moment transport equation:

$$\frac{\partial M_{i,j}}{\partial t}+\frac{\partial M_{i+1,j}}{\partial x}+\frac{\partial M_{i,j+1}}{\partial y}=i\int u^{i-1}v^{j}A_x\,n\,\mathrm{d}u\mathrm{d}v+j\int u^{i}v^{j-1}A_y\,n\,\mathrm{d}u\mathrm{d}v+\int u^{i}v^{j}\mathbb{C}\,\mathrm{d}u\mathrm{d}v$$

where the right-hand side is closed using QBMM (e.g., Gauss–Radau quadrature and CQMOM):

$$\frac{\partial M_{i,j}}{\partial t}+\frac{\partial M_{i+1,j}}{\partial x}+\frac{\partial M_{i,j+1}}{\partial y}=\sum_{\alpha=1}^{\mathcal{N}}n_\alpha\left(iu_\alpha^{i-1}v_\alpha^{j}A_{x,\alpha}+ju_\alpha^{i}v_\alpha^{j-1}A_{y,\alpha}+u_\alpha^{i}v_\alpha^{j}\mathbb{C}_\alpha\right).$$

Here $\mathcal{N}$ denotes the total number of quadrature points used in the reconstruction of $n^*(u, v)$. As described in Section 1.2, the source terms on the right-hand side are usually nonlinear and thus are closed using QBMM. On the other hand, the spatial fluxes on the left-hand side are closed in terms of moments, but some of these moments are not included in the moment set **M**.

In practice, operator splitting is used for the spatial fluxes and the source terms. Thus, we first solve

$$\frac{\partial M_{i,j}}{\partial t}+\frac{\partial M_{i+1,j}}{\partial x}+\frac{\partial M_{i,j+1}}{\partial y}=0$$

for a small time step Δt. This result is used as the initial condition to solve

$$\frac{\partial M_{i,j}}{\partial t}=\sum_{\alpha=1}^{\mathcal{N}}n_\alpha\left(iu_\alpha^{i-1}v_\alpha^{j}A_{x,\alpha}+ju_\alpha^{i}v_\alpha^{j-1}A_{y,\alpha}+u_\alpha^{i}v_\alpha^{j}\mathbb{C}_\alpha\right)$$

over the same time step for each FV cell. Note that the QBMM used for the first step (e.g., CHyQMOM) need not be the same as for the second. Besides spatial and temporal accuracy, our main concern is that after each step the moment set **M** must be realizable.

In general, when developing a numerical algorithm for the moment transport equations, there are several important things to consider:

(i) How do we discretize the spatial fluxes? (first-order, second-order, ...)

(ii) How do we update the moments in time? (explicit, implicit, ...)

(iii) How can we ensure that the moments are always realizable?

From the literature (e.g., Laurent and Nguyen, 2017; Vikas et al., 2011a), it is well known that only first-order numerical schemes for the spatial fluxes yield realizable moments. However, as mentioned above, high-order schemes are needed to reduce numerical diffusion. Getting a high-order scheme to work requires additional effort (see, e.g., Laurent and Nguyen, 2017) as we briefly describe next.

3.2 Kinetic-Based Spatial Fluxes

Given that the moments are found from a mesoscale kinetic model, kinetic-based spatial fluxes may be appropriate, making use of the reconstructed NDF to define the flux. In KBFV methods, the spatial fluxes use a kinetic formulation (Perthame, 2003). For example, with the 1-D moment advection equation defined by

$$\partial_t M_0 + \partial_x M_1 = 0,$$

we can write the spatial flux M_1 as the sum of two contributions:

$$M_1 = Q_1^- + Q_1^+ = \int_{-\infty}^{0} n^*(u)u\,\mathrm{d}u + \int_{0}^{\infty} n^*(u)u\,\mathrm{d}u.$$

The first term on the right-hand side is the contribution due to negative velocities, and the second is due to positive velocities. In the jargon of FV methods, these contributions are, respectively, the downwind and upwind flux components.

Using the reconstructed NDF

$$n^*(u) = \sum_{\alpha=1}^{N} \rho_\alpha \delta(u - u_\alpha),$$

the downwind and upwind flux components are

$$Q_1^- = \sum_{\alpha=1}^{N} \rho_\alpha u_\alpha I_{(-\infty,0)}(u_\alpha) \quad Q_1^+ = \sum_{\alpha=1}^{N} \rho_\alpha u_\alpha I_{(0,\,\infty)}(u_\alpha)$$

where $I_{\mathbb{S}}(x)$ is the indicator function for the interval $\mathbb{S}$:

$$I_{\mathbb{S}}(x) = \begin{cases} 1, x \in \mathbb{S}, \\ 0, x \notin \mathbb{S}. \end{cases}$$

The velocity abscissas thus define the fluxes locally in physical space and time. These definitions are easily extended to multivariate cases (e.g., 3-D velocity moments or velocity–mass moments), provided that the multivariate NDF is reconstructed as a finite set of weighted delta functions (Marchisio and Fox, 2013).

In the context of FV methods, the flux components are used to define the flux functions at the cell interfaces, $\mathbf{G}(\mathbf{M}_l, \mathbf{M}_r)$, given moments on the left $\mathbf{M}_l$ and right $\mathbf{M}_r$ sides of the interface (see Fig. 5). In terms of the kinetic-based fluxes, the flux function is

$$\mathbf{G}(\mathbf{M}_l, \mathbf{M}_r) = \mathbf{Q}^-(\mathbf{M}_r) + \mathbf{Q}^+(\mathbf{M}_l).$$

In words, the downwind flux carries information about the moments on the right side across the interface, while the upwind flux carries information about the moments on the left side across the interface. If we consider the interface between cell i and cell $i+1$, then the first-order flux scheme sets $\mathbf{M}_r = \mathbf{M}_{i+1}$ and $\mathbf{M}_l = \mathbf{M}_i$ where $\mathbf{M}_i$ the cell-average value in cell i. Using this method, the moments will remain realizable if the time step Δt is not too large. In contrast, a second-order flux scheme defines $\mathbf{M}_r$ and $\mathbf{M}_l$ using information from neighboring cells, e.g., $\{\mathbf{M}_{i-1}, \mathbf{M}_i, \mathbf{M}_{i+1}, \mathbf{M}_{i+2}\}$. However, there is no guarantee that this will yield realizable moment sets for any choice of Δt. As mentioned above, first-order flux functions lead to generally unacceptable levels of numerical diffusion. Thus, second-order flux functions are preferred for CFD simulations, but only if $\mathbf{M}_r$ and $\mathbf{M}_l$ can be defined such that the moments are realizable (see Chapter 8 of Marchisio and Fox, 2013, for more details).

3.3 Hyperbolic Spatial Fluxes Using HyQMOM

The kinetic-based fluxes found using the quadrature-based reconstruction of n^* are weakly hyperbolic. When HyQMOM is used, by construction, the eigenvalues of the moment system are real and distinct, and can thus be used to define the spatial fluxes (Fox et al., 2017). Let λ_i for $0 \le i \le 2N$ denote the eigenvalues found with HyQMOM. We can then define $2N+1$ weights w_i using

$$M_k = M_0 \sum_{i=0}^{2N} w_i \lambda_i^k \quad \text{for} \quad k = 0, \ldots, 2N$$

for the known set of velocity moments $\mathbf{M} = \{M_0, M_1, \ldots, M_{2N}\}$ (i.e., the moments used to compute the eigenvalues in HyQMOM). This linear system with unknown weights is well defined as long as the eigenvalues are distinct. However, there is no guarantee that the weights will be nonnegative for arbitrary N. With $N=2$, it can be shown that the weights are nonnegative (Fox et al., 2017). In fact, in general, one of the weights (say w_0) is always zero.

Assuming that the weights are nonnegative (except $w_0 = 0$) for all realizable values of $\mathbf{M}$, we can define the velocity NDF by

$$n^*(u) = M_0 \sum_{i=1}^{2N} w_i \delta(u - \lambda_i).$$

Note that this reconstruction is not a quadrature formula because the number of abscissas is not the smallest possible (i.e., $2N$ instead of $N + 1$).[c] This eigenvalue-based reconstruction of the NDF can then be used to define the kinetic-based fluxes as described in Section 3.2. In this manner, the moment system is fully hyperbolic. It should also be possible to solve the moment system using other Riemann solvers designed for hyperbolic equations (Chalons et al., 2011).

3.4 Realizable Time-Stepping Schemes

In the previous sections, we described how to define the spatial flux at a fixed time for a discrete set of FV cells. Here, we consider how to update the moment set in time for a given time step Δt. In general, we would like to control the time-integration error by using a high-order ordinary differential equation solver. In practice, either a first-order or second-order explicit scheme is usually employed in CFD codes.

Given the flux function $\mathbf{G}(\mathbf{M}_l, \mathbf{M}_r)$, we can compute the updated moments as follows (Vikas et al., 2011a):

(1) First-order explicit Euler:

$$\mathbf{M}_i^{n+1} = \mathbf{M}_i^n - \lambda\left[\mathbf{G}\left(\mathbf{M}^n_{i+\frac{1}{2},l}, \mathbf{M}^n_{i+\frac{1}{2},r}\right) - \mathbf{G}\left(\mathbf{M}^n_{i-\frac{1}{2},l}, \mathbf{M}^n_{i-\frac{1}{2},r}\right)\right]$$

(2) Second-order explicit RK2SSP:

$$\begin{aligned}
\mathbf{M}_i^{*} &= \mathbf{M}_i^n - \lambda\left[\mathbf{G}\left(\mathbf{M}^n_{i+\frac{1}{2},l}, \mathbf{M}^n_{i+\frac{1}{2},r}\right) - \mathbf{G}\left(\mathbf{M}^n_{i-\frac{1}{2},l}, \mathbf{M}^n_{i-\frac{1}{2},r}\right)\right] \\
\mathbf{M}_i^{**} &= \mathbf{M}_i^{*} - \lambda\left[\mathbf{G}\left(\mathbf{M}^{*}_{i+\frac{1}{2},l}, \mathbf{M}^{*}_{i+\frac{1}{2},r}\right) - \mathbf{G}\left(\mathbf{M}^{*}_{i-\frac{1}{2},l}, \mathbf{M}^{*}_{i-\frac{1}{2},r}\right)\right] \\
\mathbf{M}_i^{n+1} &= \frac{1}{2}\left(\mathbf{M}_i^n + \mathbf{M}_i^{**}\right)
\end{aligned}$$

where $\lambda = \Delta t/\Delta x$ is fixed using a realizability constraint, which is more restrictive than the CFL condition (Laurent and Nguyen, 2017; Vikas et al., 2011a). An important property of any scheme used for updating moments is that it must preserve realizability. In other words, if $\mathbf{M}_i^n$ is

[c] As done in Section 2.3, N is the number of nodes used when HyQMOM is degenerate, i.e., when the exact NDF is a sum of N weighted delta functions.

realizable, then $\mathbf{M}_i^{n+1}$ must also be. From the formulas above, this property obviously depends on the definition of $\mathbf{G}$. Another important fact is that the moments are convex, meaning that given realizable moments $\mathbf{M}_1$ and $\mathbf{M}_2$, the moment vector $\mathbf{M}^*$ defined by the convex sum:

$$\mathbf{M}^* = a_1\mathbf{M}_1 + a_2\mathbf{M}_2 \quad \text{with} \quad 0 \leq a_i$$

is realizable. Thus, when $\mathbf{G}$ is first order, there exists a finite λ for which the first-order explicit Euler scheme is realizable. RK2SSP is also realizable because $\mathbf{M}_i^{n+1}$ is a convex sum of two realizable moment sets. Furthermore, second-order flux functions can be defined using velocity abscissas with second-order spatial reconstruction of the weights (Laurent and Nguyen, 2017; Vikas et al., 2011a). In this way, a FV-based CFD code that is second order in space and time can be developed for solving the moment transport equations.

4. APPLICATION TO TURBULENT REACTING FLOWS

For turbulent reacting flows, probability density function (PDF) methods are very useful for modeling the concentration and temperature (or composition) statistics (Fox, 2003). As shown in Fig. 6, the concentration fields in a turbulent flow contains many different length and time scales, most of which cannot be resolved in a CFD simulation and therefore must be modeled. In a PDF model, the distribution function for the unresolved fluctuations in the concentrations provides a mesoscale description from which statistical quantities, such as the mean reaction rate, can be computed.

Details on how the PDF transport equation is derived from the microscale model are given in Fox (2003). The evolution equation of the composition PDF $f_{\boldsymbol{\phi}}(t, \mathbf{x}, \boldsymbol{\psi})$ for reactive scalars in a constant-density fluid is

$$\underbrace{\frac{\partial f_{\boldsymbol{\phi}}}{\partial t}}_{\text{rate of change}} + \underbrace{\langle U_i \rangle \frac{\partial f_{\boldsymbol{\phi}}}{\partial x_i}}_{\substack{\text{convection in physical space} \\ \text{due to the mean velocity} \\ \text{(macromixing)}}} + \underbrace{\frac{\partial}{\partial x_i}\left(\langle u_i | \boldsymbol{\psi} \rangle f_{\boldsymbol{\phi}}\right)}_{\substack{\text{convection in physical space due to} \\ \text{the scalar conditioned velocity fluctuations} \\ \text{(mesomixing)}}} =$$

$$\underbrace{-\frac{\partial}{\partial \psi_i}\left(\langle \Gamma_i \nabla^2 \phi_i' | \boldsymbol{\psi} \rangle f_{\boldsymbol{\phi}}\right) - \frac{\partial}{\partial \psi_i}\left[\left(\Gamma_i \nabla^2 \langle \phi_i \rangle\right) f_{\boldsymbol{\phi}}\right]}_{\substack{\text{transport in composition space} \\ \text{due to the molecular mixing} \\ \text{(micromixing)}}} - \underbrace{\frac{\partial}{\partial \psi_i}\left[S_i(\boldsymbol{\psi}) f_{\boldsymbol{\phi}}\right]}_{\substack{\text{transport in composition space} \\ \text{due to the chemical reactions}}}$$

where repeated Roman indices imply summation. The i-th composition variable is denoted by ϕ_i, and its phase-space component by ψ_i. Fluctuations about the mean are denoted with a prime: ϕ_i'. The velocity field is decomposed into mean and fluctuating components. As mentioned earlier, the conditional scalar flux $\langle u_i | \boldsymbol{\psi} \rangle$ can be modeled using a gradient-diffusion model. The most difficult term to model is the molecular mixing term. Remarkably, the chemical source term appears in closed form. One way to close the micromixing term is to use the IEM model introduced in Section 1.3, but it is well known that it does not give good predictions in spatially homogeneous flows (Fox, 2003). For the univariate case, a continuous PDF can be found using EQMOM (Madadi-Kandjani et al., 2017).

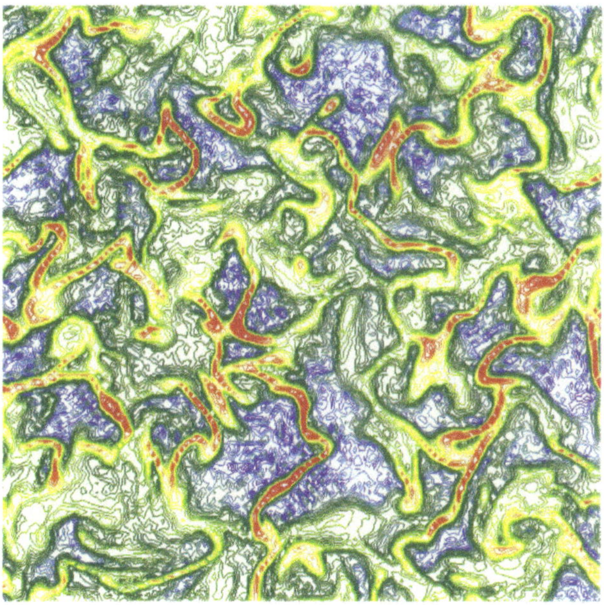

Fig. 6 A snapshot of the reaction rate for the second-order reaction $A + B \longrightarrow P$ in a non-premixed turbulent flow. The reaction rate was computed by solving the Navier–Stokes and scalar transport equations in a 3-D periodic domain with Schmidt number of 0.7. Regions in *red* correspond to the diffusion layers between A and B where the reaction rate is maximum. Regions in *blue* have essentially zero reaction rate because one of the two reactants is not present. In a macroscale CFD simulation of a chemical reactor, the 3-D domain shown in the figure would roughly correspond to a single grid cell. Thus, the observed variations in the rate of reaction would not be resolved. Instead, a PDF model for the composition probability distribution function is developed and results in a mesoscale PDF transport equation (Fox, 2003).

4.1 Univariate Case Without Reactions

For the nonreactive univariate case with $0 \leq \phi \leq 1$ (i.e., the mixture fraction), the homogeneous system with micromixing can be modeled by the Fokker–Planck (FP) model (Fox, 2003; Madadi-Kandjani et al., 2017):

$$\frac{\partial f_\phi}{\partial t} = -\frac{c_{\mathrm{FP}}+1}{2}\frac{\partial}{\partial \psi}\left[\frac{\varepsilon_\phi}{\langle \phi'^2\rangle}(\langle \phi\rangle - \psi)f_\phi\right] + \frac{c_{\mathrm{FP}}}{2}\frac{\partial^2}{\partial \psi^2}\left(\langle \epsilon_\phi|\psi\rangle f_\phi\right)$$

where the conditional scalar dissipation rate is

$$\langle \epsilon_\phi|\psi\rangle = \varepsilon_\phi \frac{\psi(1-\psi)}{\langle \phi\rangle - \langle \phi^2\rangle}.$$

The scalar dissipation rate ε_ϕ is assumed to be known (e.g., by solving a turbulence model or using $\varepsilon_\phi = C_\phi \langle \phi'^2\rangle/\tau_u$ where τ_u is the turbulence time scale). The mesoscale FP equation is closed and can thus be solved to find f_ϕ. With $c_{\mathrm{FP}} = 10$, the solution to the mesoscale PDF equation is very close to the solution to the microscale model (i.e., direct-numerical simulation (DNS) of turbulent mixing) (Madadi-Kandjani et al., 2017). However, solving the PDF equation is relatively expensive, so we would like to use moment methods.

The moment transport equations corresponding to the FP model are closed:

$$\frac{\mathrm{d}\langle \phi^m\rangle}{\mathrm{d}t} = \frac{c_{\mathrm{FP}}+1}{2}\frac{m\varepsilon_\phi}{\langle \phi'^2\rangle}\left(\langle \phi^{m-1}\rangle\langle \phi\rangle - \langle \phi^m\rangle\right) + \frac{m(m-1)c_{\mathrm{FP}}\varepsilon_\phi\left(\langle \phi^{m-1}\rangle - \langle \phi^m\rangle\right)}{2\left(\langle \phi\rangle - \langle \phi^2\rangle\right)}$$

for $m = 0, 1, \ldots, 2N$. The simplest moment closure is $N = 1$[d]:

$$\frac{\mathrm{d}\langle \phi^0\rangle}{\mathrm{d}t} = 0,$$

$$\frac{\mathrm{d}\langle \phi\rangle}{\mathrm{d}t} = 0,$$

$$\frac{\mathrm{d}\langle \phi^2\rangle}{\mathrm{d}t} = -\frac{C_\phi}{\tau_u}\langle \phi'^2\rangle.$$

When beta-EQMOM is used (Yuan et al., 2012), this closure is just the classical beta-presumed PDF model (Fox, 2003). In Fig. 7, it can be seen that the

[d] By definition, $\langle \phi^0\rangle = 1$.

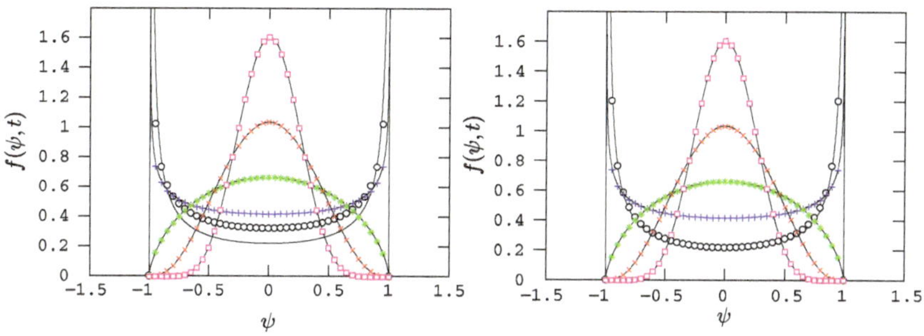

Fig. 7 Time evolution of the mixture-fraction PDF found from the mapping closure (MC) (*lines*) and beta-EQMOM (*symbols*). At time $t = 0$, the PDF is two delta functions at -1 and 1. The *left panel* is for $N = 1$ and the *right* for $N = 2$ in EQMOM. At short times, the *symbols* with $N = 1$ do not fall on MC curves (i.e., the value at $\psi = 0$ is too high). However, for $2 \leq N$, the agreement between beta-EQMOM and MC is excellent. For large times, both closures predict that the PDF will approach a Gaussian.

agreement at short times with the DNS is not satisfactory. Also, it is interesting to observe that the value of c_{FP} in the FP model has no effect on the moments up to second order.

In order to improve the agreement, $N = 2$ can be used with beta-EQMOM (see Fig. 7). In this case, two more moments are needed:

$$\frac{\mathrm{d}\langle\phi^3\rangle}{\mathrm{d}t} = \frac{3}{2}(c_{\mathrm{FP}} + 1)\frac{C_\phi}{\tau_u}\left(\langle\phi^2\rangle\langle\phi\rangle - \langle\phi^3\rangle\right) + 3c_{\mathrm{FP}}\frac{C_\phi}{\tau_u}\langle\phi'^2\rangle\frac{\left(\langle\phi^2\rangle - \langle\phi^3\rangle\right)}{\left(\langle\phi\rangle - \langle\phi^2\rangle\right)},$$

$$\frac{\mathrm{d}\langle\phi^4\rangle}{\mathrm{d}t} = 2(c_{\mathrm{FP}} + 1)\frac{C_\phi}{\tau_u}\left(\langle\phi^3\rangle\langle\phi\rangle - \langle\phi^4\rangle\right) + 6c_{\mathrm{FP}}\frac{C_\phi}{\tau_u}\langle\phi'^2\rangle\frac{\left(\langle\phi^3\rangle - \langle\phi^4\rangle\right)}{\left(\langle\phi\rangle - \langle\phi^2\rangle\right)}.$$

Note that these moment equations are closed, depend on c_{FP}, and can be solved with any ODE solver for initial boundary problems. As shown in Madadi-Kandjani et al. (2017), this fourth-order moment closure using the beta-EQMOM reconstruction gives very accurate predictions for the mixture-fraction PDF for binary mixing at all times. In fact, increasing N beyond 2 shows no further improvement. Moreover, the solutions for $\langle\phi^3\rangle$ and $\langle\phi^4\rangle$ reach self-similar forms very quickly for large c_{FP}, which would suggest that the algebraic models found by setting the right-hand sides of the ODEs for $\langle\phi^3\rangle$ and $\langle\phi^4\rangle$ to zero and $c_{\mathrm{FP}} = \infty$ can be used to simplify the closure. In other words, it is not necessary to solve transport equations for $\langle\phi^3\rangle$ and $\langle\phi^4\rangle$ for binary mixing problems (e.g., for use with the flamelet model or with the conditional moment closure).

4.2 Multivariate Case With Chemical Reactions

For the multivariate case (i.e., more than one composition variable), the reconstruction of the PDF is more complicated. Moreover, when the number of composition variables is large, the number of low-order moments quickly becomes intractable to reproduce (Fox, 2009a). Physically, however, all of the moments are not independent because they are coupled by the chemical reactions. It often suffices to choose one variable, say ϕ_1, and work with the conditional moments of the other components. For nonpremixed systems, ϕ_1 is normally the mixture fraction (Fox, 2003). Nonetheless, ϕ_1 can be any variable that spans the range of variability in the system (e.g., a progress variable in a premixed system such as temperature, etc.).

Here, we will discuss the case where $\phi_1 = \psi_1$ is the conditioning variable, and ϕ_i with $i \in 2, \ldots, M$ are the remaining components. In order to reconstruct the PDF of ϕ_1, we need the moments $\langle \phi^k \rangle$ for $k = 0, 1, 2, 3, 4$. (see Section 4.1). Using these five moments, a Gauss–Radau quadrature can be used to reconstruct the marginal PDF of ϕ_1[e]:

$$f_1(\psi_1) = \sum_{\alpha=0}^{2} p_\alpha \delta(\psi_1 - \psi_{1,\alpha}) \implies \langle \phi_1^k \rangle = \sum_{\alpha=0}^{2} p_\alpha \psi_{1,\alpha}^k$$

where $\psi_{1,0}$ is a known lower bound for ϕ_1 (e.g., $\psi_{1,0} = 0$ for mixture fraction). The Gauss–Radau quadrature generates probabilities $0 < p_0, p_1, p_2$ and abscissas $(\psi_{1,1}, \psi_{1,2})$ from the moments $(1, \langle \phi_1 \rangle, \langle \phi_1^2 \rangle, \langle \phi_1^3 \rangle, \langle \phi_1^4 \rangle)$.[f]

For the other components, we define the conditional moments by

$$\langle \phi_i^j | \psi_1 \rangle = \int \psi_i^j f_i(\psi_i | \psi_1) \, \mathrm{d}\psi_i$$

where $f_i(\phi_i | \psi_1)$ is the conditional PDF of ϕ_i given $\phi_1 = \psi_1$. These conditional moments are related to the unconditional bivariate moments by

$$\langle \phi_1^k \phi_i^j \rangle = \int \psi_1^k \langle \phi_i^j | \psi_1 \rangle f_1(\psi_1) \, \mathrm{d}\psi_1 = \sum_{\alpha=0}^{2} p_\alpha \psi_{1,\alpha}^k \langle \phi_i^j | \psi_{1,\alpha} \rangle.$$

---

[e] As discussed in Section 4.1, the same moments can be used to find a continuous representation of f_1 using beta-EQMOM. In either case, the moments of ϕ_1 up to fourth order are exactly reproduced.

[f] In the singular cases where $\phi_0 = 0$ or $\phi_1 = 1$ for mixture fraction (i.e., in the unmixed feed streams), two of the probabilities will be null.

In words, $\langle \phi_i^j | \psi_1 \rangle$ is a function of ψ_1 defined such that it yields the bivariate moments. The final form on the right-hand side was found by evaluating the integral using the Gauss–Radau reconstruction of f_1. In practice, a set of bivariate moments $\langle \phi_1^k \phi_i^j \rangle$ defined by the ranges of j, k will be known, and we must find a suitable functional form for $\langle \phi_i^j | \psi_1 \rangle$. This can be done using CQMOM (Yuan and Fox, 2011).

In order to illustrate the moment-inversion procedure with CQMOM, we consider the first-order conditional moment $\langle \phi_i | \psi_1 \rangle$ and write it as a polynomial in ψ_1[g]:

$$\langle \phi_i | \psi_1 \rangle = a_{i,0} + a_{i,1}\psi_1 + a_{i,2}\psi_1^2$$

with three unknown coefficients $(a_{i,0}, a_{i,1}, a_{i,2})$. Injecting this expression into the definition of the bivariate moments yields

$$\langle \phi_1^k \phi_i \rangle = \sum_{\alpha=0}^{2} p_\alpha \psi_{1,\alpha}^k (a_{i,0} + a_{i,1}\psi_{1,\alpha} + a_{i,2}\psi_{1,\alpha}^2) = a_{i,0}\langle \phi_1^k \rangle + a_{i,1}\langle \phi_1^{k+1} \rangle + a_{i,2}\langle \phi_1^{k+2} \rangle.$$

The right-hand side is linear in the unknown coefficients. We can thus define a linear system by taking $k = 0, 1, 2$:

$$\begin{bmatrix} 1 & \langle \phi_1 \rangle & \langle \phi_1^2 \rangle \\ \langle \phi_1 \rangle & \langle \phi_1^2 \rangle & \langle \phi_1^3 \rangle \\ \langle \phi_1^2 \rangle & \langle \phi_1^3 \rangle & \langle \phi_1^4 \rangle \end{bmatrix} \begin{bmatrix} a_{i,0} \\ a_{i,1} \\ a_{i,2} \end{bmatrix} = \begin{bmatrix} \langle \phi_i \rangle \\ \langle \phi_1 \phi_i \rangle \\ \langle \phi_1^2 \phi_i \rangle \end{bmatrix}.$$

Here, the system matrix is the Hankel matrix for moments up to fourth order in ϕ_1 (Dette and Studden, 1997):

$$\mathbf{H}_4 = \begin{bmatrix} 1 & \langle \phi_1 \rangle & \langle \phi_1^2 \rangle \\ \langle \phi_1 \rangle & \langle \phi_1^2 \rangle & \langle \phi_1^3 \rangle \\ \langle \phi_1^2 \rangle & \langle \phi_1^3 \rangle & \langle \phi_1^4 \rangle \end{bmatrix}.$$

If the moments are realizable, then $0 \leq |\mathbf{H}_4|$ (the determinant of the Hankel matrix must be nonnegative). The linear system will have a unique solution if the determinant is positive.

[g] Other functional forms, such as piecewise cubic Hermite interpolating polynomials, can be used. However, the number of unknowns must be consistent with order of the Gauss–Radau quadrature employed for f_1.

From the theory of moments (Dette and Studden, 1997), it is known that $0 = |\mathbf{H}_4|$ corresponds to one of two cases:

(i) The variance of ϕ_1 is null (i.e., $0 = |\mathbf{H}_2|$).

(ii) The probability $p_0 = 0$ in the Gauss–Radau quadrature for f_1 (i.e., $0 < |\mathbf{H}_2|$).

For the first, we set $a_{i,\,1} = a_{i,\,2} = 0$ so that $a_{i,\,0} = \langle\phi_i\rangle$. For the second, we set $a_{i,\,2} = 0$ and find

$$\begin{bmatrix} a_{i,0} \\ a_{i,1} \end{bmatrix} = \frac{1}{\langle\phi_1'^2\rangle}\begin{bmatrix} \langle\phi_1^2\rangle\langle\phi_i\rangle - \langle\phi_1\rangle\langle\phi_1\phi_i\rangle \\ \langle\phi_1\phi_i\rangle - \langle\phi_1\rangle\langle\phi_i\rangle \end{bmatrix}.$$

Otherwise, the full system can be solved for $(a_{i,\,0}, a_{i,\,1}, a_{i,\,2})$.

In summary, given the composition moments

$$\mathbf{\Phi} = \{\langle\phi_1\rangle, \langle\phi_1^2\rangle, \langle\phi_1^3\rangle, \langle\phi_1^4\rangle, \langle\phi_i\rangle, \langle\phi_1\phi_i\rangle, \langle\phi_1^2\phi_i\rangle \mid \forall i \in 2, 3, \ldots, M\};$$

we can reconstruct the marginal PDF $f_1(\psi_1)$ and the conditional moments $\langle\phi_i|\psi_1\rangle = a_{i,0} + a_{i,1}\psi_1 + a_{i,2}\psi_1^2$. Hereinafter, we will describe the case with $M = 2$, for which the transported moment set contains seven moments:

$$\mathbf{\Phi} = \{\langle\phi_1\rangle, \langle\phi_2\rangle, \langle\phi_1^2\rangle, \langle\phi_1\phi_2\rangle, \langle\phi_1^3\rangle, \langle\phi_1^2\phi_2\rangle, \langle\phi_1^4\rangle\}.$$

The extension to more components is straightforward. Each additional component (i.e., $2 < M$) adds three moments to the definition of $\mathbf{\Phi}$ and, thus, the overall computational cost increases like $3M$. Although it will not be described here, it is possible to go to higher-order conditional moments such as $\langle\phi_i^2|\psi_1\rangle$ using CQMOM (Yuan and Fox, 2011).

4.3 Algorithm for Turbulent Reacting Flows

The moment transport equations are found from the *closed*[h] mesoscale PDF transport equation as described in Section 1.3. For the moments, the most important unclosed term is the chemical source term for composition ϕ_i, denoted by $\langle\phi_1^j S_i(\boldsymbol{\phi})\rangle$, and defined by

$$\langle\phi_1^j S_i(\boldsymbol{\phi})\rangle = \int \psi_1^j S_i(\boldsymbol{\psi}) f_{\boldsymbol{\phi}}(\boldsymbol{\psi})\,\mathrm{d}\boldsymbol{\psi}.$$

In order to approximate this integral, the joint composition PDF is reconstructed using the Gauss–Radau quadrature and the conditional moments:

[h] In other words, the scalar flux and micromixing models have already been introduced so that the PDF transport equation can be solved for the joint composition PDF (Fox, 2003).

$$f_{\phi}(\psi_1,\psi_2)=\sum_{\alpha=0}^{2}p_\alpha\delta(\psi_1-\psi_{1,\alpha})\delta(\psi_2-a_{2,0}-a_{2,1}\psi_{1,\alpha}-a_{2,2}\psi_{1,\alpha}^2).$$

The QBMM approximation for the chemical source term is then

$$\langle\phi_1^j S_i(\boldsymbol{\phi})\rangle=\sum_{\alpha=0}^{2}p_\alpha\psi_{1,\alpha}^j S_i(\psi_{1,\alpha},\psi_{2,\alpha})$$

where $\psi_{2,\alpha}=a_{2,0}+a_{2,1}\psi_{1,\alpha}+a_{2,2}\psi_{1,\alpha}^2$. The extension to more than two compositions is obvious.

In practice, the moment transport equations are solved in a CFD code by using operator splitting. The overall algorithm has the following steps:

Step 1. Initialize the composition moments $\mathbf{\Phi}(0, \mathbf{x})$ in the computational domain.

Step 2. Solve scalar transport equations for advection and diffusion of the composition moments using realizable FV schemes.

Step 3. Update composition moments for the micromixing source term in each grid cell using a realizable ODE solver.

Step 4. Update composition moments for the chemical source term in each grid cell using a realizable (stiff) ODE solver.

Step 5. Advance in time by repeating from *Step 2* until the simulation is complete.

The chemical source term in *Step 4* does not change the probabilities p_α. Thus, to ensure composition moment realizability, it is convenient to write *Step 4* in terms of the abscissas:

$$\frac{\partial\psi_{1,\alpha}}{\partial t}=S_1(\psi_{1,\alpha},\psi_{2,\alpha}),\quad \frac{\partial\psi_{2,\alpha}}{\partial t}=S_2(\psi_{1,\alpha},\psi_{2,\alpha}).$$

These coupled ODEs are solved separately for each $\alpha = 0, 1, 2$ and for each grid cell. Thus, after *Step 3*, the composition moments are used to find (p_α, $\psi_{1,\alpha}$, $\psi_{2,\alpha}$). Then, in *Step 4*, ($\psi_{1,\alpha}$, $\psi_{2,\alpha}$) are updated due to chemistry. *Step 4* ends by computing the updated composition moments using $\langle\phi_1^j\phi_2\rangle=\sum_{\alpha=0}^{2}p_\alpha\psi_{1,\alpha}^j\psi_{2,\alpha}$. Note that the resulting composition moments will be physically[i] realizable as long as the chemistry solver provides physically realistic values for the abscissas.

Reacting-flow models based on a delta function representation have a long history in chemical reaction engineering and are often called "multienvironment mixing" models. Such models were first written is a

[i] Because the p_α are nonnegative, the composition moments will always be statistically realizable.

Lagrangian framework, but can be directly related to CFD models based on PDF methods (see Fox, 1998, for details). As explained in Fox (2003), the differences between the multienvironment models proposed by different researchers are essentially how the micromixing is described. It is also important to not confuse multienvironment models and compartmental (or zone) models. In the latter, spatial averaging over many CFD cells is invoked to reduce the computational cost of the solving the model. In fact, because the spatial resolution is low, a compartmental model should include a multienvironment model to account for unresolved concentration fluctuations, which are always present in a turbulent flow.

5. APPLICATION TO THE POPULATION BALANCE EQUATION FOR FINE PARTICLES

Quadrature-based moment methods were first used for aerosols undergoing aggregation (McGraw, 1997). For most fine-particle applications (Raman and Fox, 2016), the particles follow the fluid phase with essentially zero diffusivity. However, due to aggregation, the particles form clusters of different sizes. Thus, in order to account for the number of particles of a given mass ξ, an NDF is introduced. Similar to chemical species (e.g., polymer molecules of different lengths), the NDF $n(t, \mathbf{x}, \xi)$ obeys a PBE of the form

$$\frac{\partial n(\xi)}{\partial t} + \mathbf{U}_{\mathrm{g}} \cdot \nabla n(\xi) = \int_0^{\xi} \beta(\xi', \xi - \xi') n(\xi') n(\xi - \xi') \, \mathrm{d}\xi' - n(\xi) \int_0^{\infty} \beta(\xi, \xi') n(\xi') \, \mathrm{d}\xi'$$

where $\mathbf{U}_{\mathrm{g}}(t, \mathbf{x})$ is the gas velocity found by solving the Navier–Stokes equation. $\beta(\xi, \xi')$ is the (mass-dependent) aggregation kernel, which determines the rate at which particles of masses ξ and ξ' aggregate. The integral terms on the right-hand side couple together all of the different sizes, making the analytical solution for $n(\xi)$ intractable. Moreover, in most applications, the gas velocity field will be turbulent, further complicating the task.

When other physics, such as nucleation, growth, and breakage, are included in the PBE, the integral terms can be quite complicated (Cheng and Fox, 2010; Cheng et al., 2010; Marchisio et al., 2003a,b,c; Piton et al., 2000; Sung et al., 2011, 2014; Wick et al., 2017; Zucca et al., 2006). Also, often knowledge of just the particle size is insufficient so a multivariate PBE is needs for $n(t, \mathbf{x}, \boldsymbol{\xi})$ with multiple internal variables $\boldsymbol{\xi}$. QBMM provides a tractable, yet very accurate, way to solve a PBE in the context of CFD simulations. Below, we first consider the case of laminar flow. Then, by combining moment methods with multienvironment models for turbulent flows, we briefly consider the CFD model for fine-particle formation in turbulent flows (Raman and Fox, 2016).

5.1 Fine Particles in Laminar Flows

In a laminar flow, $\mathbf{U}_g(t, \mathbf{x})$ can be taken as known (e.g., found in a separate CFD simulation). For simplicity, we assume that $n(\xi)$ does not change the fluid velocity and consider only aggregation. For QBMM, we need transport equations for the integer moment $M_k(t,\mathbf{x}) = \int n(t,\mathbf{x},\xi)\xi^k d\xi$. Using the PBE above, it is straightforward to show (Marchisio and Fox, 2013) that the moment transport equation is

$$\frac{\partial M_k}{\partial t} + \mathbf{U}_g \cdot \nabla M_k = \frac{1}{2}\sum_{\alpha=0}^{N}\sum_{\gamma=0}^{N} n_\alpha n_\gamma \beta(\xi_\alpha, \xi_\gamma)\left[(\xi_\alpha + \xi_\gamma)^k - \xi_\alpha^k - \xi_\gamma^k\right].$$

This system of equations must be solved for $k \in 0, 1, \ldots, 2N$. The Gauss–Radau quadrature is applied to the moments $\mathbf{M}$ to find the $N+1$ weights n_α and N unknown abscissas ξ_α. As described in Section 2.2, ξ_0 is equal to the smallest particle size. A typical initial condition is that none of the particles are clustered: $M_0(0, \mathbf{x})$ is the prescribed number concentration field and $M_k(0,\mathbf{x}) = M_0(0,\mathbf{x})\xi_0^k$. In terms of the weights and abscissas, this initial condition corresponds to $n_0(0, \mathbf{x}) = M_0(0, \mathbf{x})$ and $n_1(0, \mathbf{x}) = n_2(0, \mathbf{x}) = 0$. Note that these initial conditions correspond to a delta function NDF, which cannot easily be handled by moment methods other that QBMM (Nguyen et al., 2017). Such "degenerate" NDF arise in many real applications due to the underlying physics. A robust CFD code for fine particles must therefore be able to deal with degenerate NDF without user intervention.

In practice, the CFD code will employ operator splitting in order to ensure realizability of the moments at each step. In the first step, a realizable FV scheme is used to solve the moment advection equation (Laurent and Nguyen, 2017):

$$\frac{\partial M_k}{\partial t} + \mathbf{U}_g \cdot \nabla M_k = 0.$$

At the end of this step, the updated moments are then used to calculate the NDF weights and abscissas. Then, a realizable ODE solver is used to solve for the change in the moments due to aggregation:

$$\frac{\partial M_k}{\partial t} = S_k(\mathbf{M}) = \frac{1}{2}\sum_{\alpha=0}^{N}\sum_{\gamma=0}^{N} n_\alpha n_\gamma\, \beta(\xi_\alpha, \xi_\gamma)\left[(\xi_\alpha + \xi_\gamma)^k - \xi_\alpha^k - \xi_\gamma^k\right]$$

in each grid cell. Here, $S_k(\mathbf{M})$ is the "reaction rate" for moment M_k. When other physics (e.g., nucleation, growth) are added, dedicated realizable solvers should be employed for each contribution (see Nguyen et al., 2017, for details).

5.2 Fine Particles in Turbulent Flows

At first glance, the modeling of fine particles in turbulent flows can seem quite complicated because one must deal with a PDF transport equation for turbulent mixing and the PBE for the particle NDF (Piton et al., 2000; Wang and Fox, 2003a,b). However, if one keeps in mind that the PBE describes the laminar flow, as in the previous section, then it is simple to associate the moments M_k with the components in the composition vector $\boldsymbol{\phi}$ in the PDF formulation described in Section 4. Specifically, for a non-premixed turbulent flow, we let ϕ_1 be the mixture fraction, and

$$\phi_2 = M_0, \phi_3 = M_1, \phi_4 = M_2, \ldots, \phi_{2N+1} = M_{2N}$$

where N is the number of quadrature nodes used in the Gauss–Radau quadrature for the NDF. In words, the NDF moments are the components of the composition vector. Thus, in each environment, there will be a separate set of moments: $\boldsymbol{\psi}_\alpha = \{\psi_{1,\alpha}, M_{0,\alpha}, \ldots, M_{2N,\alpha}\}$. The "reaction rate" for the composition moment set in each environment is $S_k(\mathbf{M}_\alpha)$ with $\mathbf{M}_\alpha = \{M_{0,\alpha}, \ldots, M_{2N,\alpha}\}$.

Specifically, the composition moments needed for a CFD simulation of fine particles in a turbulent flow are

$$\boldsymbol{\Phi} = \{\langle\phi_1\rangle, \langle\phi_1^2\rangle, \langle\phi_1^3\rangle, \langle\phi_1^4\rangle, \langle M_k\rangle, \langle\phi_1 M_k\rangle, \langle\phi_1^2 M_k\rangle \mid \forall k \in 0, 1, \ldots, 2N\}.$$

Given these moments, Gauss–Radau quadrature and CQMOM yield the environment weights and compositions:

$$\{p_0, p_1, p_2\}, \{\phi_{1,0}, \phi_{1,1}, \phi_{1,2}\}, \{M_{k,0}, M_{k,1}, M_{k,2} \mid \forall k \in 0, 1, \ldots, 2N\}.$$

Finally, Gauss–Radau quadrature applied to the NDF moments in each environment produces the NDF weights and abscissas:

$$\{M_{0,\alpha}, M_{1,\alpha}, \ldots, M_{2N,\alpha}\} \Longrightarrow \{n_{0,\alpha}, n_{1,\alpha}, \ldots, n_{N,\alpha}\}, \{\xi_{0,\alpha}, \xi_{1,\alpha}, \ldots, \xi_{N,\alpha}\}.$$

The latter are used in the "reaction solver" to update the NDF weights and abscissas due to aggregation.

The extension of the algorithm presented above to turbulent reacting flows with fine-particle production follows the same logic (Raman and Fox, 2016). The principal difference is that the composition vector in the PDF transport equation will include chemical species and (multivariate) NDF moments. However, from a computational standpoint, each substep in the algorithm is done in the same way as described above (see Mehta et al., 2010, 2013; Sung et al., 2011, 2014, for specific examples). The CFD code should thus be structured into modules that handle each step in a systematic fashion (Passalacqua et al., 2018).

6. APPLICATION TO THE KINETIC EQUATION FOR GAS–PARTICLE FLOWS

As an alternative to Eulerian moment methods, the KE can be simulated using Lagrangian methods where individual particles are tracked, including particle–particle collisions (Capecelatro and Desjardins, 2013). While quite accurate, these methods are very expensive for moderately dense flows where the number of particles in the system can be many millions, or even billions. The question then arises: Do Eulerian moment methods and Lagrangian methods give equivalent results?

Here, we use cluster-induced turbulence (CIT) (Capecelatro et al., 2015b) to investigate this question for gas–particle flows. The simulation cases considered are as follows. They correspond to Cases 1–6 in Capecelatro et al. (2015a) (see Fig. 8).

(i) Triply periodic domain with gravity pointing downward.
(ii) Terminal (Stokes) velocity of particles: $\boldsymbol{\nu} = 0.1$ m/s.
(iii) Particle Reynolds number: $\mathrm{Re}_\mathrm{p} = \frac{d_\mathrm{p} \nu}{\nu_\mathrm{g}} = 0.5$.
(iv) Average volume fraction of particles: $\alpha_\mathrm{p} = 0.01$.
(v) Solid–gas density ratio: $\rho_\mathrm{p}/\rho_\mathrm{g} = 1000$.
(vi) Characteristic cluster length scale: $\mathcal{L} = 2.5$ mm.
(vii) Box size for largest case: $L_x/\mathcal{L} = 129$ $(2048 \times 512 \times 512)$.

The Lagrangian and Eulerian simulations are performed on the same grid for the fluid phase, but not with same numerical schemes. In general, the

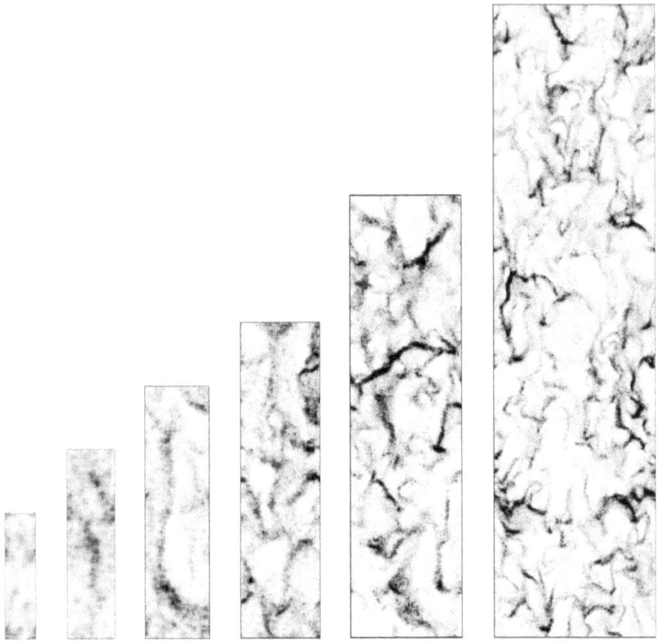

Fig. 8 2-D cuts through the 3-D periodic domains used in the six Lagrangian simulations of cluster-induced turbulence (CIT). The domains are shown to scale relative to the largest domain. The location of the particles is shown, which is roughly equal to the particle-phase volume fraction when scaled properly. On the largest domain (Case 6), the clusters are smaller than the box size, while on the smaller domains some clusters span the entire width of the box. The flow statistics depend on the domain size, but the dependence is relative weak for Case 6. Note that the particles are nonadhesive and hence the clusters freely form and breakup due to the fluid forces. The mean fluid velocity is null, and the presence of falling clusters disturbs the fluid, causing turbulence.

Lagrangian simulations use high-order schemes and thus differences at the smallest resolved scales are expected. Moreover, the Lagrangian method uses a smoothing filter to transfer information from the particles to the fluid (Capecelatro and Desjardins, 2013). The computational grids are cubic with length approximately equal to the particle diameter.

The Eulerian simulations use a second-order velocity moment closure with an anisotropic Gaussian (AG) distribution (i.e., a total of 10 velocity moments) (Kong et al., 2017; Vié et al., 2015). A first-order, kinetic-based flux is used for the free-transport terms in the moment transport equations. The moment equations have been implemented in OpenFOAM. The governing equation for the Eulerian 10-moment system (valid for dilute flows) is

$$\frac{\partial \mathbf{M}}{\partial t} + \nabla \cdot \mathbf{F} = \mathbf{A} + \mathbf{C} \quad \text{where} \quad M_{ijk} = \int u^i v^j w^k f(\boldsymbol{v}) \mathrm{d}\boldsymbol{v}.$$

The 10 moments are related to the particle-phase variables as follows:

$$M^0_{000} = \alpha_\mathrm{p}, \quad \begin{bmatrix} M^1_{100} \\ M^1_{010} \\ M^1_{001} \end{bmatrix} = \alpha_\mathrm{p} \mathbf{U}_\mathrm{p}, \quad \begin{bmatrix} M^2_{200} & M^2_{110} & M^2_{101} \\ M^2_{110} & M^2_{020} & M^2_{011} \\ M^2_{101} & M^2_{011} & M^2_{002} \end{bmatrix} = \alpha_\mathrm{p} (\mathbf{U}_\mathrm{p} \otimes \mathbf{U}_\mathrm{p} + \mathbf{P}_\mathrm{p})$$

where $\mathbf{U}_\mathrm{p}$ is particle-phase velocity, $\mathbf{P}_\mathrm{p}$ is the (symmetric) particle-phase pressure tensor. One-third the trace of $\mathbf{P}_\mathrm{p}$ is equal to the granular temperature Θ_p. In the KBFV code, a Gauss–Hermite quadrature is used to solve for the free-transport flux $\mathbf{F}$ (Kong et al., 2017). As the AG closure is strictly hyperbolic (Vié et al., 2015), no delta-shocks are present. However, particle trajectory crossing cannot be captured with the AG closure. The collision term $\mathbf{C}$ is approximated by the inelastic BGK model (Passalacqua et al., 2011). As in the Lagrangian simulation, Stokes drag and gravity are used to define $\mathbf{A}$.

A qualitative comparison of clustering is shown in Fig. 9 for Case 6. While the cluster shapes are very similar, the Eulerian simulation has slightly longer/wider clusters. A quantitative comparison of selected one-point statistics is shown in Figs. 10 and 11. The moment method (EE-AG) has slightly fewer high α_p values than the Lagrangian (EL) simulations. This difference is likely due to numerical resolution of the smallest scales. EE-AG has slightly higher falling velocity for the particles, which is likely due to the larger clusters. In Fig. 11, EE-AG has lower granular temperature than EL, which is partially due to how it is extracted from the EL data in post-processing. On the other hand, EE-AG has higher turbulent kinetic energy for both phases, which is likely due to the larger clusters. In general, the domain-size dependence found with EL and EE-AG is very similar, thus either method can be used to investigate the effect of domain size. Quantitative comparisons between two-point statistics such as energy spectra confirm the qualitative observations about the cluster size discussed above (Patel et al., 2017).

6.1 Governing Equations for All Gas–Particle Flow Regimes

The CIT results above are for dilute flows where only the free-transport flux is important in the Eulerian model. For dense flows, such as fluidized beds, the collisional and frictional fluxes are also very important. The numerical treatment of such fluxes is not particularly efficient using kinetic-based

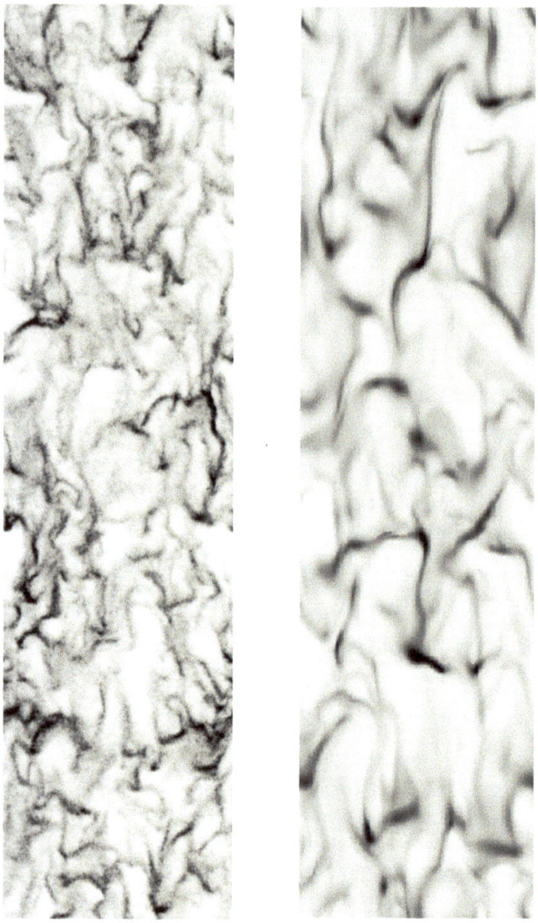

Fig. 9 Particle volume fraction fields for (*left*) Lagrangian and (*right*) Eulerian simulations from Case 6. In general, clusters on the left are larger than on the right. It is found that the Eulerian solution depends more strongly on the grid refinement, and the clusters tend to get smaller as the Eulerian grid is refined further (Patel et al., 2017).

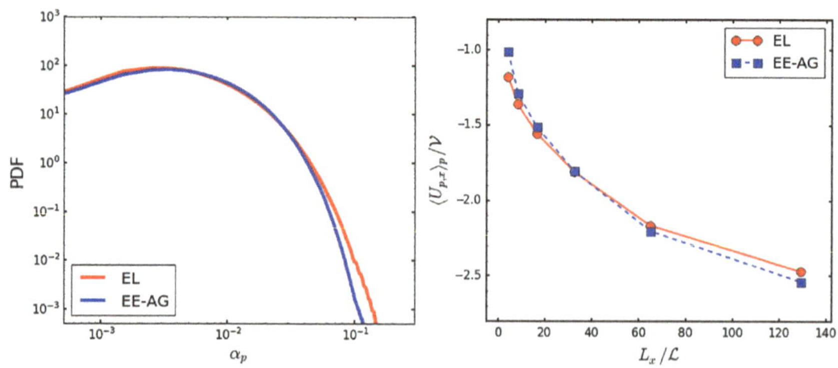

Fig. 10 Particle volume fraction PDF for Case 6 (*left*) and falling velocity (*right*) for all cases. The magnitude of the falling velocity is larger for larger domains because larger clusters can be resolved, which fall faster. In all cases, the agreement between EE-AG and EL is reasonably good.

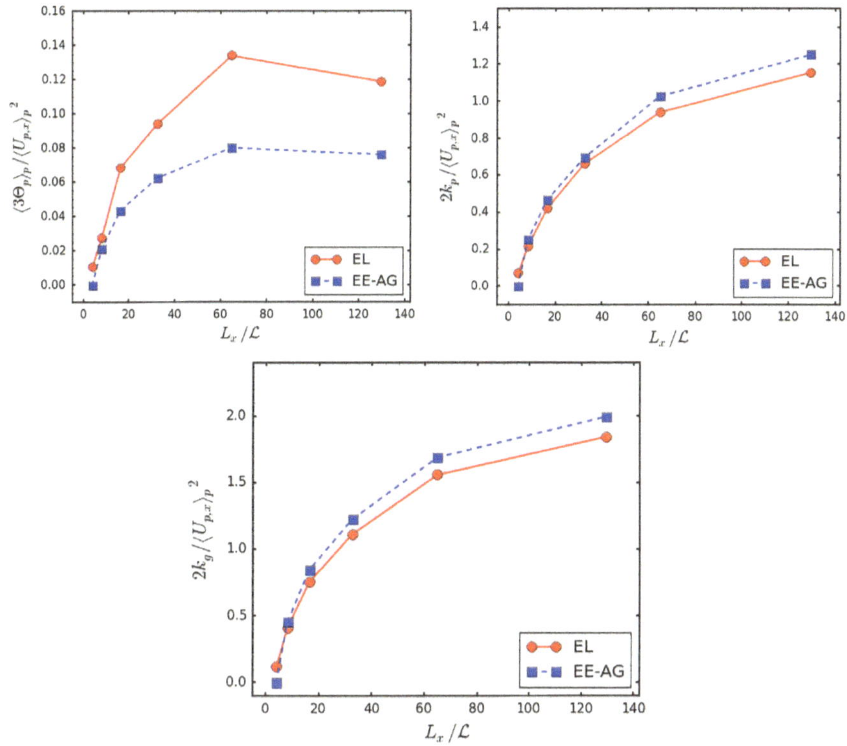

Fig. 11 Kinetic energy statistics for granular temperature Θ_p (*top left*), turbulent kinetic energy in the particle-phase k_p (*top right*) and in the fluid-phase k_f (*bottom*) for all cases. The agreement between the two simulation methods improves as the grid is further refined (Patel et al., 2017), but some difference persist due to the closures inherent in the EE model.

methods. Thus, we can combine KBFV methods for the free-transport flux with classical hydrodynamic models for the dense phase in order to treat all flow regimes in gas–particle flows (Kong and Fox, 2017). This algorithm is briefly described in the next section. Here, we provide an overview of the governing equations and point out the terms that are dominant in each flow regime.

The gas–particle flow model for dilute-to-dense particle flows couples two models. The first is the (monodisperse) particle-phase KE

$$\frac{\partial n}{\partial t} + \mathbf{v} \cdot \frac{\partial n}{\partial \mathbf{x}} + \frac{\partial}{\partial \mathbf{v}} \cdot \mathbf{A} n = \mathbb{C}$$

where $n(t,\mathbf{x},\mathbf{v})$ is the velocity NDF, $\mathbf{v}$ is the particle velocity, $\mathbf{A}$ is the particle acceleration, and $\mathbb{C}$ it the rate of change of n due to Boltzmann–Enskog collisions and frictional stresses. The corresponding gas-phase equations are

$$\frac{\partial}{\partial t}\alpha_{\mathrm{g}}\rho_{\mathrm{g}} + \nabla\cdot\alpha_{\mathrm{g}}\rho_{\mathrm{g}}\mathbf{U}_{\mathrm{g}} = 0$$
$$\frac{\partial}{\partial t}\alpha_{\mathrm{g}}\rho_{\mathrm{g}}\mathbf{U}_{\mathrm{g}} + \nabla\cdot\alpha_{\mathrm{g}}\rho_{\mathrm{g}}\mathbf{U}_{\mathrm{g}}\mathbf{U}_{\mathrm{g}} = \nabla\cdot\alpha_{\mathrm{g}}\boldsymbol{\tau}_{\mathrm{g}} + \boldsymbol{\beta}_{\mathrm{g}} + \alpha_{\mathrm{g}}\rho_{\mathrm{g}}\mathbf{g}$$

where $\alpha_{\mathrm{g}} = 1 - \alpha_{\mathrm{p}}$ is the gas-phase volume fraction, β_{g} is the mean particle drag. In dense flows, the frictional stresses arising from $\mathbb{C}$ make the KE very stiff so that only the five lowest-order velocity moments used in the hydrodynamic model (i.e., α_{p}, $\mathbf{U}_{\mathrm{p}}$, Θ_{p}) are needed to describe the particle phase. In contrast, for dilute flows, at least the 10 AG moments are needed to approximate the nonequilibrium behavior in the particle phase. The computational algorithm must explicitly account for these differences when solving for the spatial fluxes.

The moment transport equations with collisional and frictional fluxes can be written for the 10 lowest-order velocity moments (i.e., up to second order). In the dilute regime, we have as before

$$\frac{\partial\mathbf{M}}{\partial t} + \nabla\cdot\mathbf{F} = \mathbf{S}$$

where

$$M^0_{000} = \rho_{\mathrm{p}}\alpha_{\mathrm{p}}, \quad \begin{bmatrix} M^1_{100} \\ M^1_{010} \\ M^1_{001} \end{bmatrix} = \rho_{\mathrm{p}}\alpha_{\mathrm{p}}\mathbf{U}_{\mathrm{p}}, \quad \begin{bmatrix} M^2_{200} & M^2_{110} & M^2_{101} \\ M^2_{110} & M^2_{020} & M^2_{011} \\ M^2_{101} & M^2_{011} & M^2_{002} \end{bmatrix} = \rho_{\mathrm{p}}\alpha_{\mathrm{p}}(\mathbf{U}_{\mathrm{p}}\otimes\mathbf{U}_{\mathrm{p}} + \mathbf{P}_{\mathrm{p}}).$$

$\mathbf{F}$ is the free-transport flux and $\mathbf{S}$ is the source term due to acceleration and collisions (Fox and Vedula, 2010). In the dense regime, the governing equations for the particle-phase mass, momentum, and pressure tensor are

$$\frac{\partial\rho_{\mathrm{p}}\alpha_{\mathrm{p}}}{\partial t} + \nabla\cdot\rho_{\mathrm{p}}\alpha_{\mathrm{p}}\mathbf{U}_{\mathrm{p}} = 0$$
$$\frac{\partial\rho_{\mathrm{p}}\alpha_{\mathrm{p}}\mathbf{U}_{\mathrm{p}}}{\partial t} + \nabla\cdot\rho_{\mathrm{p}}\alpha_{\mathrm{p}}\left(\mathbf{U}_{\mathrm{p}}\otimes\mathbf{U}_{\mathrm{p}} + \mathbf{P}_{\mathrm{p}} + \mathbf{G}_{\mathrm{p}} + \mathbf{Z}_{\mathrm{p}}\right) = \rho_{\mathrm{p}}\alpha_{\mathrm{p}}\boldsymbol{g} + \rho_{\mathrm{p}}\alpha_{\mathrm{p}}\boldsymbol{M}_{\mathrm{pg}}$$
$$\frac{\partial\rho_{\mathrm{p}}\alpha_{\mathrm{p}}\mathbf{P}_{\mathrm{p}}}{\partial t} + \nabla\cdot\rho_{\mathrm{p}}\alpha_{\mathrm{p}}\left(\mathbf{U}_{\mathrm{p}}\otimes\mathbf{P}_{\mathrm{p}} + \mathbf{Q}_{\mathrm{p}} + \mathbf{H}_{\mathrm{p}}\right)$$
$$+\rho_{\mathrm{p}}\alpha_{\mathrm{p}}\left[(\mathbf{P}_{\mathrm{p}} + \mathbf{G}_{\mathrm{p}})\cdot\nabla\mathbf{U}_{\mathrm{p}} + (\nabla\mathbf{U}_{\mathrm{p}})^T\cdot(\mathbf{P}_{\mathrm{p}} + \mathbf{G}_{\mathrm{p}})\right] = \rho_{\mathrm{p}}\alpha_{\mathrm{p}}\mathbf{E}_{\mathrm{pg}} + \rho_{\mathrm{p}}\alpha_{\mathrm{p}}\mathbf{C}_{\mathrm{p}}$$

where $trace(\mathbf{P}_p) = 3\Theta_p$ and the free-transport fluxes are shown in blue. The other fluxes correspond to the collisional and frictional contributions, which are dominant in the dense regime.

In the dense regime, the pressure tensor in the free-transport flux can be approximated by the hydrodynamic model shown here in red:

$$\mathbf{P}_p = \Theta_p \mathbf{I} - \boldsymbol{\sigma}_p = \Theta_p \mathbf{I} - 2\nu_{p,k} \boldsymbol{S}_p$$

where the particle-phase rate-of-deformation tensor is

$$\boldsymbol{S}_p = \frac{1}{2}\left[\nabla \mathbf{U}_p + (\nabla \mathbf{U}_p)^T - \frac{2}{3}(\nabla \cdot \mathbf{U}_p)\mathbf{I}\right],$$

and $\nu_{p,k}$ is an effective kinematic viscosity for the particle phase. The collisional flux (where the corresponding pressure is infinite for finite $\alpha_p \approx 0.63$) is modeled as

$$\mathbf{G}_p = \frac{p_{p,c}}{\rho_p \alpha_p}\mathbf{I} - 2\nu_{p,c} \boldsymbol{S}_p.$$

The frictional flux (where the corresponding pressure is infinite for finite $\alpha_p \approx 0.63$, but null when $\alpha_p < 0.55$) is modeled as

$$\mathbf{Z}_p = \frac{p_{p,f}}{\rho_p \alpha_p}\mathbf{I} - 2\nu_{p,f} \boldsymbol{S}_p.$$

In the hydrodynamic model, the energy fluxes are modeled by $\mathbf{U}_p \otimes \mathbf{P}_p + \mathbf{Q}_p + \mathbf{H}_p = \mathbf{U}_p \otimes \mathbf{P}_p - \frac{2}{3} k_\Theta \nabla \otimes \mathbf{P}_p$, where k_Θ is an effective conductivity. In the dense regime, the transport equation for Θ_p is solved and the traceless, anisotropic part of $\mathbf{P}_p$ (i.e., $\boldsymbol{\sigma}_p$) is modeled in terms of $\boldsymbol{S}_p$.

6.2 Solution Algorithm for All Gas–Particle Flow Regimes

In the kinetic flux-splitting scheme for all flow regimes, the free-transport flux is treated with moment methods in dilute regions and with the hydrodynamic model in dense regions. This is accomplished as follows. First, we split the free-transport fluxes $\mathbf{F}$ into two contributions in the moment transport equation:

$$\frac{\partial \mathbf{M}}{\partial t} + \nabla \cdot h_1 \mathbf{F} + \nabla \cdot (h_2 \mathbf{F} + \mathbf{G} + \mathbf{Z}) = \mathbf{S}$$

where $h_1 + h_2 = 1$ and the flux-splitting function is defined by using the kinetic, collisional, and frictional contributions to the particle-phase pressure:

$$h_2 := \left(\frac{p_{\mathrm{p,c}} + p_{\mathrm{p,f}}}{p_{\mathrm{p,k}} + p_{\mathrm{p,c}} + p_{\mathrm{p,f}} + \varepsilon} \right)^p .$$

Note that this definition of h_2 is not unique. In general, h_2 should be unity in regions where the frictional flux is nonnegligible, and zero elsewhere. For example, a smooth function of α_{p} that changes rapidly from zero to one near $\alpha_{\mathrm{p}} = 0.5$ may be suitable for this purpose.

The blue flux in the moment transport equation above is treated using KBFV methods, while the red flux uses the hydrodynamic model. Recall that KBFV methods are explicit in time, making it very difficult to set a hard upper limit on α_{p} in the dense regime. Physically, the speed of sound in the particle phase approaches infinity as α_{p} approaches α_{max}, causing the time step in an explicit scheme to approach zero. In dilute regions, h_2 is nearly zero, while in dense regions it is nearly one (see Fig. 12). Thus, the exponent p determines the fraction of the free-transport flux that is treated using (the more accurate, but explicit) KBFV methods. Experience suggests the $p = 4$ is a reasonable choice (Kong and Fox, 2017). In a CFD simulation, h_2 will be different in every grid cell at every time step so that the CFD code dynamically adjusts to the *local* flow regime.

The operator-splitting algorithm used to solve the moment transport equation proceeds in two steps:

Step (i) KBFV method:

$$\frac{\partial \mathbf{M}}{\partial t} + \boldsymbol{\nabla} \cdot h_1 \mathbf{F} = \mathbf{0}$$

where quadrature-based moment methods are used together with the kinetic-based fluxes described in Section 3.2 to advance the 10 moments in **M**.

Step (i) Hydrodynamic solver:

$$\frac{\partial \mathbf{M}}{\partial t} + \boldsymbol{\nabla} \cdot (h_2 \mathbf{F} + \mathbf{G} + \mathbf{Z}) = \mathbf{S}$$

where (realizable) implicit FV methods are used that can deal with the strongly nonlinear dependence of the frictional stresses on the particle-phase volume fraction. In this step, the moment vector **M**

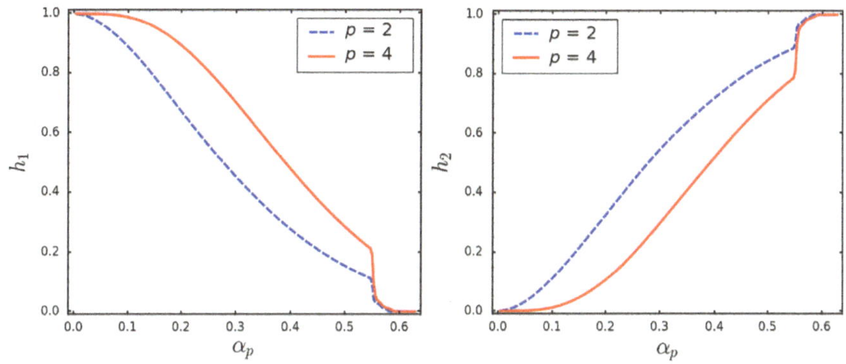

Fig. 12 Dependency of flux-splitting functions on α_p found from a fluidized-bed simulation. Below 10%, the free-transport flux is treated using kinetic-based methods. Above 55% the implicit hydrodynamic solver is used. In between, a linear combination of the two methods is used to treat free transport. (The collisional and frictional fluxes are treated exclusively using the implicit hydrodynamic solver.) Other definitions are also possible as long as h_2 is very near unity in the frictional regime, and very near zero in dilute flows.

is split into the five hydrodynamic moments $\{\alpha_p, \mathbf{U}_p, \Theta_p\}$ and the deviatoric part $\boldsymbol{\sigma}_p$. The former is advanced in the two-fluid hydrodynamic solver, while the latter is advanced as a passive scalar field. At the end of this step, $\mathbf{M}$ is recomputed based on its components (i.e., from the definition of the central moments).

Note that all of the source terms $\mathbf{S}$ are evaluated in the *Step (ii)*. Here, $\mathbf{G}$ and $\mathbf{Z}$ are the collisional and frictional fluxes, respectively, for the velocity moments $\mathbf{M}$, which depend only on $\{\alpha_p, \mathbf{U}_p, \Theta_p\}$. The choice of separating the two steps as done above was made to allow the use of any existing two-fluid solver for gas–particle flows to be used in *Step (ii)*. In the results shown below, the OpenFOAM two-fluid solver is used for this step.

In summary, the overall solution algorithm for treating all gas–particle flow regimes uses the following steps:

Step 1. Initialize all variables $\mathbf{M}$, $\{\alpha_p, \mathbf{U}_p, \Theta_p, \boldsymbol{\sigma}_p\}$, and $\{\alpha_g, \mathbf{U}_g, p_g\}$.

Step 2. Calculate h_1 and h_2.

Step 3. Using the *explicit* free-transport solver:

Compute kinetic-based moment fluxes to transport the moments.

Update $\{\alpha_p, \mathbf{U}_p, \Theta_p, \boldsymbol{\sigma}_p\}$ using the updated moments $\mathbf{M}$.

Step 4. Using the *iterative* hydrodynamic solver:

Solve hydrodynamic transport equations for $\{\alpha_p, \mathbf{U}_p, \Theta_p\}$.

Solve gas-phase velocity and pressure $\{\mathbf{U}_g, p_g\}$ equations.

Step 5. Solve scalar transport equation for $\boldsymbol{\sigma}_p$.

Step 6. Update 10 moments $\mathbf{M}$ using the updated $\{\alpha_p, \mathbf{U}_p, \Theta_p, \boldsymbol{\sigma}_p\}$.

Step 7. Advance in time by repeating from *Step 2* until the simulation is complete.

In practice, the computational cost is dominated by the hydrodynamic solver when the flow is dense (e.g., a fluidized bed). The algorithm can be implemented with any hydrodynamic solver (e.g., MFIX, OpenFOAM) developed for gas–particle flows. The only substantial change needed in the hydrodynamic code is to insert h_2 in the definition of the free-transport flux term. However, care must be taken to ensure that the numerical algorithm remains conservative (Passalacqua and Fox, 2011a).

6.3 Application to Particle-Laden Flows

The splitting method described in the previous section was developed mainly to extend moment methods to dense regimes, such as fluidized beds. In a bubbling fluidized bed, dense regions coexist with very dilute regions (i.e., bubbles). In Fig. 13, snapshots of the particle volume fraction and h_2 fields are shown for a fluidized-bed simulation. From the value of h_2, it is evident than very dilute (h_2 near zero) and very dense (h_2 near one) regions exist throughout the fluidized bed. The curves in Fig. 12 were computed by time/space averaging the value of h_2 for a given value of α_p from the fluidized-bed simulation with different values of p.

Results for a 3-D wall-bounded vertical channel, simulated with the algorithm described above, are shown in Fig. 14. Three channel-average particle volume fractions were employed: $\overline{\alpha}_p = 0.01$, 0.1, and 0.4, corresponding to dilute, moderately dilute and moderately dense flows. Simulations were also run with $h_2 = 1$ for all cases (i.e., the hydrodynamic model). For all values of $\overline{\alpha}_p$, the level of granular temperature was significantly larger when the free-transport flux was treated with KBFV methods. Analysis of the results (Kong and Fox, 2017) revealed that the source term in the hydrodynamic model for the granular temperature was too small by a factor of approximately two. In general, the flow dynamics of gas–particle flows are significantly different when the granular temperature is lower. For example, the particle volume fraction distribution across the channel width changes significantly. In dense flows, the observed bubble size is larger when the free-transport flux is treated using a moment method (see Kong and Fox, 2017, for examples).

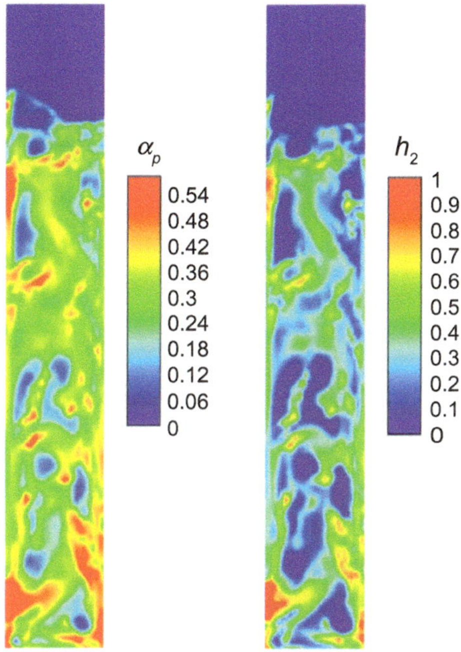

Fig. 13 Instantaneous results for a fluidized-bed simulation using the algorithm described in Section 6.2. The value of h_2 is the fraction of the free-transport flux treated by the hydrodynamic closure. It is evident that h_2 and α_p are strongly correlated as expected. Note that even in a dense fluidized bed, void zones (bubbles) are treated with kinetic-based finite-volume methods.

Current (as yet unpublished) work on particle-laden channel flows is focused on the following points:

(i) Verification of Eulerien–Eulerian simulations against Eulerian–Lagrangian simulations over a wide range of fluid-phase Reynolds numbers, Stokes numbers and mass loadings.

(ii) Investigation of the abrupt transition from fully developed channel flow to CIT as the overall mass loading is increased.

(iii) Study of how the ratio of the channel width to cluster size changes the observed flow regimes, and how this ratio alters the scale-up rules for risers.

In additional, further investigations are needed for nonisothermal and/or polydisperse systems in order to gain a complete understanding of the momentum and energy transport in these fascinating multiphase flows.

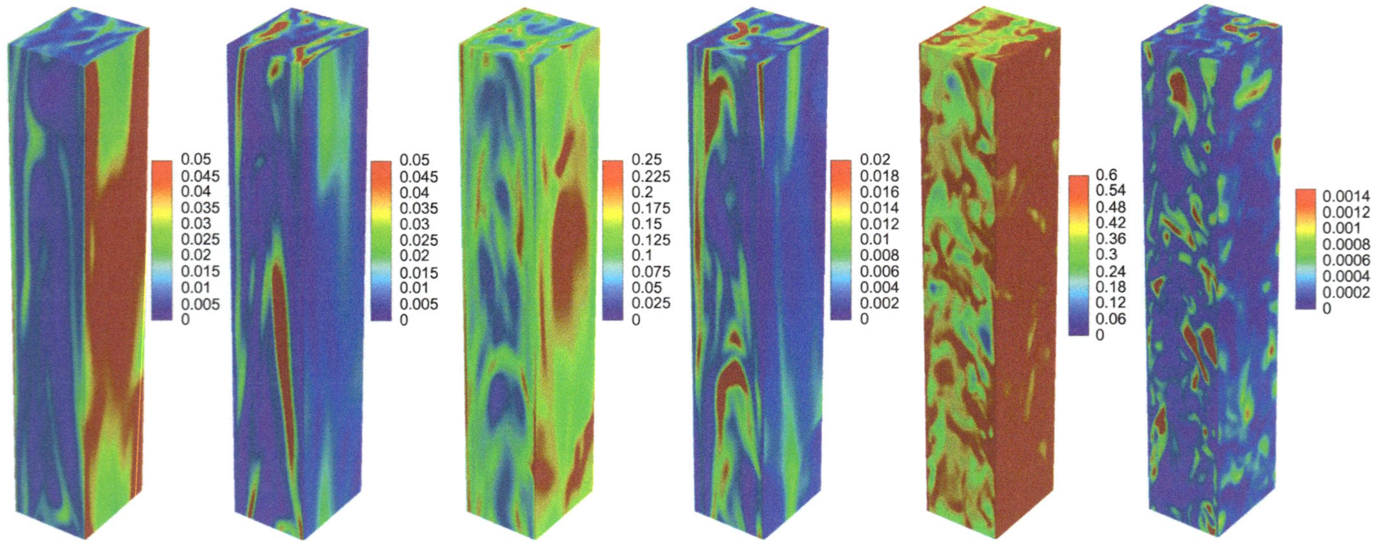

Fig. 14 Particle volume fraction (*left*) and granular temperature (*right*) fields found from the free-transport flux-splitting code for vertical channel flow. *Left panels*: Dilute with $\overline{\alpha}_p = 0.01$. *Center panels*: Moderately dilute with $\overline{\alpha}_p = 0.1$. *Right panels*: Moderately dense with $\overline{\alpha}_p = 0.4$. The overall computational time is mainly controlled by the iterative hydrodynamic solver, especially for the moderately dense flow. In the dilute flow, particles are mainly falling as clusters near the channel walls. In the moderately dense flow, clusters span the entire channel width.

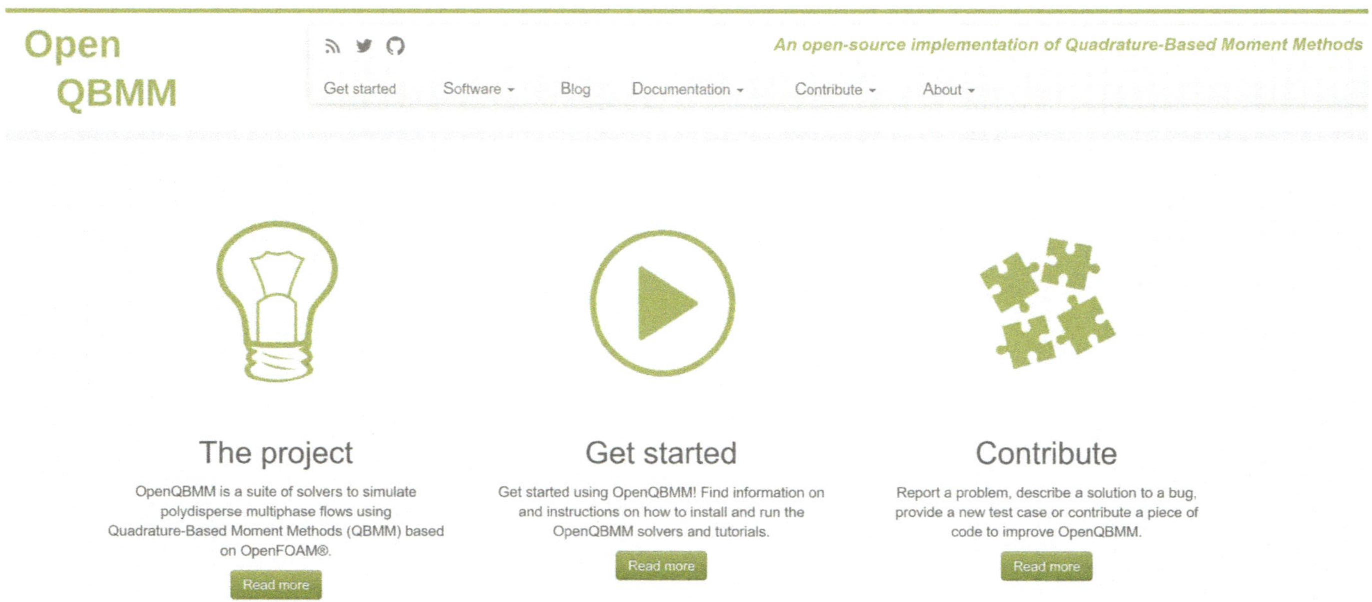

Fig. 15 Quadrature-based moment methods have been implemented in OpenFOAM as part of the OpenQBMM project (http://www.openqbmm.org/). The current release has tools for fine particles, bubbly flows, and monodisperse gas–particle flows (Passalacqua et al., 2018). Future releases will include polydisperse gas–particle and chemically reacting flows. Researchers interested in applying these methods to their own problems are encouraged to contribute and become members of the OpenQBMM community.

7. CONCLUDING REMARKS

In this chapter, a brief overview of recent developments in quadrature-based moment methods applied to multiphase and reacting flows has been presented. As noted in Section 1 the focus is on new developments occuring after the publication of the monograph of Marchisio and Fox (2013). Based on our current knowledge, we can make the following concluding remarks.

(a) Mesoscale models have a direct link with the underlying physics and result in a PBE/KE/GPBE for the PDF/NDF.

(b) Quadrature-based moment methods solve the PBE/KE/GPBE by reconstructing the PDF/NDF from its lower-order moments.

(c) When solving the KE/GPBE for dilute fluid–particle flows, it is best to use the velocity NDF reconstruction and kinetic-based finite-volume methods.

(d) When solving the KE/GPBE for dense fluid–particle flows, it is best to use a kinetic-based finite-volume method for the free-transport flux and a two-fluid hydrodynamic solver for the collisional and frictional fluxes.

(e) The joint mass–velocity NDF allows for polydisperse, inertial particles needed in applications involving particle "size" changes.

(f) High-order finite-volume methods are needed to minimize "numerical diffusion," but such methods must be designed to be realizable.

(g) The QBMM algorithms described in this chapter are part of the open-source OpenQBMM project (see Fig. 15 for more details).

The interested reader can find all of the necessary details in the original publications. A semi-complete list of papers on this topic by the author and his coworkers is given in References. A more extensive reference list of related papers is provided in Marchisio and Fox (2013).

REFERENCES

Capecelatro J, Desjardins O: An Euler–Lagrange strategy for simulating particle-laden flows, *J Comput Phys* 238:1–31, 2013.

Capecelatro J, Desjardins O, Fox RO: Effect of domain size on fluid-particle dynamics in homogeneous, gravity-driven, cluster-induced turbulence, *J Fluids Eng* 138(4):041301, 2015a.

Capecelatro J, Desjardins O, Fox RO: On fluid-particle dynamics in fully developed cluster-induced turbulence, *J Fluid Mech* 780:578–635, 2015b.

Chalons C, Fox RO, Massot M: A multi-Gaussian quadrature method of moments for gas-particle flows in a LES framework. In *Proceedings of the summer program 2010*, Stanford, 2010a, Center for Turbulence Research, pp 347–358.

Chalons C, Kah D, Massot M: Beyond pressureless gas dynamics: quadrature-based velocity moment methods, *Commun Math Sci* 10:1241–1272, 2010b.
Chalons C, Fox RO, Laurent F, Massot M, Vié A: A multi-Gaussian quadrature method of moments for simulating high-stokes-number turbulent two-phase flows. In *Annual research briefs, 2011*, Stanford, 2011, Center for Turbulence Research, pp 309–320.
Chalons C, Fox RO, Laurent F, Massot M, Vié A: Multivariate Gaussian extended quadrature method of moments for turbulent disperse multiphase flow, *Multiscale Model Simul* 15(4):1553–1583, 2017.
Cheng JC, Fox RO: Kinetic modeling of nanoprecipitation using CFD coupled with a population balance, *Ind Eng Chem Res* 49:10651–10662, 2010.
Cheng JC, Vigil RD, Fox RO: A competitive aggregation model for Flash NanoPrecipitation, *J Colloid Interface Sci* 351:330–342, 2010.
de Chaisemartin S, Fréret L, Kah D, et al: Turbulent combustion of polydisperse evaporating sprays with droplet crossing: Eulerian modeling and validation in the infinite Knudsen limit. In *Proceedings of the summer program 2008*, Stanford, 2008, Center for Turbulence Research, pp 265–276.
de Chaisemartin S, Fréret L, Kah D, et al: Eulerian models for turbulent spray combustion with polydispersity and droplet crossing, *CR Mec* 337(6):438–448, 2009.
Dette H, Studden WJ: *The theory of canonical moments with applications in statistics, probability, and analysis*, New York, 1997, John Wiley and Sons.
Fan R, Marchisio DL, Fox RO: Application of the direct quadrature method of moments to polydisperse gas-solid fluidized beds, *Powder Technol* 139(1):7–20, 2004.
Fox RO: On the relationship between Lagrangian micromixing models and computational fluid dynamics, *Chem Eng Process: Process Intesif* 37(6):521–535, 1998.
Fox RO: *Computational models for turbulent reacting flows*, UK, 2003, Cambridge University Press.
Fox RO: Bivariate direct quadrature method of moments for coagulation and sintering of particle populations, *J Aerosol Sci* 37(11):1562–1580, 2006.
Fox RO: Optimal moment sets for the multivariate direct quadrature method of moments, *Ind Eng Chem Res* 48:9686–9696, 2009a.
Fox RO: On multiphase turbulence models for collisional fluid-particle flows, *J Fluid Mech* 742:368–424, 2014.
Fox RO, Laurent F, Massot M: Numerical simulation of spray coalescence in an eulerian framework: direct quadrature method of moments and multi-fluid method, *J Comput Phys* 227(6):3058–3088, 2008.
Fox RO, Vedula P: Quadrature-based moment model for moderately dense polydisperse gas-particle flows, *Ind Eng Chem Res* 49:5174–5187, 2010.
Fox RO, Laurent F, Vié A: Conditional hyperbolic quadrature method of moments for kinetic equations, *J Comput Phys*, 2017, under review. https://hal.archives-ouvertes.fr/hal-01632813.
Gautschi W: *Orthogonal polynomials: computational and approximation*, UK, 2004, Oxford University Press.
Kah D, Laurent F, Fréret L, et al: Eulerian quadrature-based moment models for dilute polydisperse evaporating sprays, *Flow Turbul Combust* 85(3):649–676, 2010.
Kong B, Fox RO: A solution algorithm for fluid-particle flows across all flow regimes, *J Comput Phys* 344:575–594, 2017.
Kong B, Fox RO, Feng H, Capecelatro J, Patel R, Desjardins O: Euler-euler anisotropic gaussian mesoscale simulation of homogeneous cluster-induced gas-particle turbulence, *AIChE J* 63:2630–2643, 2017.
Laurent F, Nguyen TT: Realizable high-order finite-volume schemes for the advection of moment sets of the particle size distribution, *J Comput Phys* 337:309–338, 2017.
Madadi-Kandjani E, Fox RO, Passalacqua A: Application of the Fokker-Planck molecular mixing model to turbulent scalar mixing using moment methods, *Phys Fluids* 29:065109, 2017.

Marchisio DL, Fox RO: Solution of population balance equations using the direct quadrature method of moments, *J Aerosol Sci* 36(1):43–73, 2005.

Marchisio DL, Fox RO: *Computational models for polydisperse particulate and multiphase systems,* UK, 2013, Cambridge University Press.

Marchisio DL, Pikturna JT, Fox RO, Vigil RD, Barresi AA: Quadrature method of moments for population-balance equations, *AIChE J* 49(5):1266–1276, 2003.

Marchisio DL, Vigil RD, Fox RO: Implementation of the quadrature method of moments in CFD codes for aggregation-breakage problems, *Chem Eng Sci* 58(15):3337–3351, 2003.

Marchisio DL, Vigil RD, Fox RO: Quadrature method of moments for aggregation-breakage processes, *J Colloid Interface Sci* 258(2):322–334, 2003.

McGraw R: Description of aerosol dynamics by the quadrature method of moments, *Aerosp Sci Technol* 27:255–265, 1997.

Mehta M, Sung Y, Raman V, Fox RO: Multiscale modeling of TiO_2 nanoparticle production in flame reactors: effect of chemical mechanism, *Ind Eng Chem Res* 49:10663–10673, 2010.

Mehta M, Raman V, Fox RO: On the role of gas-phase and surface chemistry in the production of titania nanoparticles in turbulent flames, *Chem Eng Sci* 104:1003–1018, 2013.

Nguyen TT, Laurent F, Fox RO, Massot M: Solution of population balance equations in applications with fine particles: mathematical modeling and numerical schemes, *J Comput Phys* 325:129–156, 2017.

Passalacqua A, Fox RO: An iterative solution procedure for multi-fluid gas-particle flow models on unstructured grids, *Powder Technol* 213:174–187, 2011a.

Passalacqua A, Galvin JE, Vedula P, Hrenya CM, Fox RO: A quadrature-based kinetic model for dilute non-isothermal granular flows, *Commun Comput Phys* 10(1):216–252, 2011.

Passalacqua A, Laurent F, Madadi-Kandjani E, Heylmum J, Fox RO: An open-source quadrature-based population balance solver for OpenFOAM, *Chem Eng Sci* 176:306–318, 2018.

Patel R, Desjardins O, Kong B, Capecelatro J, Fox RO: Verification of Eulerian-Eulerian and Eulerian-Lagrangian simulations for turbulent fluid-particle flows, *AIChE J* 63:1–17, 2017(online).

Perthame B: *Kinetic formulation of conservation laws,* UK, 2003, Oxford University Press.

Piton D, Fox RO, Marcant B: Simulation of fine particle formation by precipitation using computational fluid dynamics, *Can J Chem Eng* 78(5):983–993, 2000.

Raman V, Fox RO: Modeling of fine-particle formation in turbulent flames, *Annu Rev Fluid Mech* 48:159–190, 2016.

Sung Y, Raman V, Fox RO: Large-eddy simulation based multiscale modeling of TiO_2 nanoparticle synthesis in turbulent flame reactors using detailed nucleation chemistry, *Chem Eng Sci* 66(19):4370–4381, 2011.

Sung Y, Raman V, Koo H, Mehta M, Fox RO: Large-eddy simulation modeling of turbulent flame synthesis of titania nanoparticles using a bivariate particle description, *AIChE J* 60(2):459–472, 2014.

Vié A, Doisneau F, Massot M: On the anisotropic Gaussian velocity closure for inertial-particle laden flows, *Commun Comput Phys* 17(1):1–46.

Vikas V, Wang ZJ, Passalacqua A, Fox RO: Realizable high-order finite-volume schemes for quadrature-based moment methods, *J Comput Phys* 230(13):5328–5352, 2011.

Vikas V, Wang ZJ, Fox RO: Realizable high-order finite-volume schemes for quadrature-based moment methods applied to diffusion population balance equations, *J Comput Phys* 249:162–179, 2013.

Wang L, Fox RO: Comparison of micromixing models for CFD simulation of nanoparticle formation, *AIChE J* 50(9):2217–2232, 2003a.

Wang L, Fox RO: Application of in situ adaptive tabulation to CFD simulation of nanoparticle formation by reactive precipitation, *Chem Eng Sci* 58(19):4387–4401, 2003b.
Wick A, Nguyen TT, Laurent F, Fox RO, Pitsch H: Modeling soot oxidation with the extended quadrature method of moments, *Proc Combust Inst* 36:789–797, 2017.
Yuan C, Fox RO: Conditional quadrature method of moments for kinetic equations, *J Comput Phys* 230(22):8216–8246, 2011.
Yuan C, Laurent F, Fox RO: An extended quadrature method of moments for population balance equations, *J Aerosol Sci* 51:1–23, 2012.
Zucca A, Marchisio DL, Barresi AA, Fox RO: Implementation of the population balance equation in CFD codes for modelling soot formation in turbulent flames, *Chem Eng Sci* 61(1):87–95, 2006.

FURTHER READING

Buffo A, Vanni M, Marchisio DL, Fox RO: Multivariate quadrature-based moment methods for turbulent polydisperse gas-liquid systems, *Int J Multiphase Flow* 50:41–57, 2013.
Capecelatro J, Desjardins O, Fox RO: Numerical study of collisional particle dynamics in cluster-induced turbulence, *J Fluid Mech* 747:1–10, 2014.
Capecelatro J, Desjardins O, Fox RO: Strongly coupled fluid-particle flows in vertical channels. I. Reynolds-averaged two-phase turbulence statistics, *Phys Fluids* 28:033306, 2016.
Capecelatro J, Desjardins O, Fox RO: Strongly coupled fluid-particle flows in vertical channels. II. Turbulence modeling, *Phys Fluids* 28:033307, 2016.
Desjardins O, Fox RO, Villedieu P: A quadrature-based moment closure for the Williams spray equation. In *Proceedings of the summer program, 2006,* Stanford, 2006, Center for Turbulence Research, pp 223–234.
Desjardins O, Fox RO, Villedieu P: A quadrature-based moment method for dilute fluid-particle flows, *J Comput Phys* 227(4):2514–2539, 2008.
Emre O, Fox RO, Massot M, de Chaisemartin S, Jay S, Laurent F: Towards Eulerian modeling of a polydisperse evaporating spray under realistic internal-combustion-engine conditions, *Flow Turbul Combust* 93(4):689–722, 2014.
Emre O, Kah D, Jay S: Eulerian moment methods for automotive sprays, *Atomization Sprays* 25(3):189–254, 2015.
Fox RO: A quadrature-based third-order moment method for dilute gas-particle flows, *J Comput Phys* 227(12):6313–6350, 2008.
Fox RO: Higher-order quadrature-based moment methods for kinetic equations, *J Comput Phys* 228(20):7771–7791, 2009b.
Fox RO: Large-eddy-simulation tools for multiphase flows, *Annu Rev Fluid Mech* 44:47–76, 2012.
Fox RO: *Quadrature-based moment methods for polydisperse multiphase flows,* Vienna, 2013, Stochastic Methods in Fluid Mechanics, pp 87–136.
Fox RO: Turbulence in multiphase flows. In Yeoh G, editor: *Handbook of multiphase flow science and technology*, Singapore, 2016, Springer, https://doi.org/10.1007/978-981-4585-86-6_2-1.
Icardi M, Asinari P, Marchisio DL, Izquierdo S, Fox RO: Quadrature-based moment closures for non-equilibrium flows: hard-sphere collisions and approach to equilibrium, *J Comput Phys* 231(21):7431–7449, 2012.
Madadi-Kandjani E, Passalacqua A: An extended quadrature-based moment method with log-normal kernel density function, *Chem Eng Sci* 131:323–339, 2015.
Passalacqua A, Fox RO: Advanced continuum modeling of gas-particle flows beyond the hydrodynamic limit, *Appl Math Model* 35(4):1616–1627, 2011b.

Passalacqua A, Fox RO: Simulation of mono- and bidisperse gas-particle flow in a riser with a third-order quadrature-based moment method, *Ind Eng Chem Res* 52:187–198, 2013.
Passalacqua A, Fox RO, Garg R, Subramaniam S: A fully coupled quadrature-based moment method for dilute to moderately dilute fluid-particle flows, *Chem Eng Sci* 65(7):2267–2283, 2010.
Vikas V, Hauck CD, Wang ZJ, Fox RO: Radiation transport modeling using extended quadrature methods of moments, *J Comput Phys* 246:221–241, 2013.
Vikas V, Yuan C, Wang ZJ, Fox RO: Modeling of bubble-column flows with quadrature-based moment methods, *Chem Eng Sci* 66(14):3058–3070, 2011.
Yuan C, Kong B, Passalacqua A, Fox RO: An extended quadrature-based mass-velocity moment model for polydisperse bubbly flows, *Can J Chem Eng* 92(4):689–722, 2014.

CHAPTER TWO

Numerical Simulation of Multiphase Reactive Flows

Ziad Hamidouche*, Enrica Masi*,[1], Pascal Fede*, Renaud Ansart†, Hervé Neau*, Mehrdji Hemati†, Olivier Simonin*

*Institut de Mécanique des Fluides de Toulouse, IMFT, Université de Toulouse, CNRS-Toulouse, France
†Laboratoire de Génie Chimique, LGC, Université de Toulouse, CNRS-Toulouse, France
[1]Corresponding author: e-mail address: enrica.masi@imft.fr

Contents

1. Introduction 55
2. Brief Description of the Experimental Campaign 57
3. Modeling Description 59
   3.1 Eulerian Approach 60
   3.2 Methane–Air Premixed Combustion Modeling 61
   3.3 Thermal Radiation Model 63
   3.4 Local Phase Temperature Determination 64
4. Numerical Simulations 65
   4.1 Code Description and Performances 65
   4.2 Reactive Fluidized Bed Simulations 72
5. Results and Discussions 75
   5.1 Preliminary Results 77
   5.2 Effect of Heat Exchanges at the Wall 79
   5.3 Effect of the Grid Refinement 81
   5.4 Mesoscopic and Macroscopic Scale Analysis 87
   5.5 Evaluation of the Subgrid Turbulent Mixing Effect on the Combustion Process 104
6. Conclusion 108
Acknowledgments 109
Appendix. Modeling Reactive Gas–Particle Flows 109
   A.1 Gas-Phase Modeling 110
   A.2 Dispersed Phase Modeling 113
   A.3 Closure Laws 114
References 121

Abstract

This study deals with mathematical modeling and numerical simulations of reactive multiphase flows in dense fluidized beds. These flows involve complex physical mechanisms related to the coupling between the bed hydrodynamic and the reactions,

Advances in Chemical Engineering, Volume 52
ISSN 0065-2377
https://doi.org/10.1016/bs.ache.2018.01.003

which are still poorly understood. In this context, numerical simulations can provide explanatory access to the underlying physics taking place in the reactor, thus supplementing the experimental results. The present contribution focuses on the natural gas combustion in a dense fluidized bed reactor, for which experimental results are available from the literature (Dounit, 2001; Dounit et al., 2001, 2008). In their experiments, the authors pointed out the essential role played by the particle projection zone, above the bed surface, in the global thermal efficiency of the reactor operating at relatively low temperatures (600°C–800°C). In the present study, this point is further investigated by analyzing the results obtained by the numerical simulations. The unsteady 3D numerical simulations were performed using NEPTUNE_CFD code which is based on an Euler–Euler approach; the latter computes both the gas and the particulate phases in an Eulerian framework, accounting for specific closures modeling the interphase momentum and energy transfers. Time-averaged quantities were then computed and compared with the available experimental measurements. Numerical results (especially the gas temperature) were found to be very sensitive to the mesh refinement for the selected operating point. A further analysis at mesoscopic and macroscopic scales was carried out. This analysis pointed out the crucial role of the toroidal loop, which extends from the bed, close to the ejection zone, to the freeboard, on the fuel conversion. In numerical simulations, this loop must be accurately reproduced in order to provide reliable combustion predictions.

NOMENCLATURE

LATIN SYMBOLS

C_D drag coefficient (–)
C_{p_k} mass heat capacity of the phase k (J kg^{-1} K^{-1})
D_p^{kin} particle diffusivity due to the particle agitation (m^2 s^{-1})
D_{gp}^t dispersion coefficient (m^2 s^{-1})
D_α^{eff} effective species diffusivity (m^2 s^{-1})
D_α^l laminar species diffusivity (m^2 s^{-1})
D_α^t turbulent species diffusivity (m^2 s^{-1})
d_p particle diameter (m)
e_c particle–particle restitution coefficient (–)
$F_{r,i}$ drag force i-component (N)
g_i gravity acceleration i-component (m^2 s^{-1})
g_0 radial function (–)
H_α mass enthalpy of the species α (J kg^{-1})
H_k mass enthalpy of the phase k (J kg^{-1})
H_σ interface mass enthalpy (J kg^{-1})
h_k instantaneous mass enthalpy of the phase k (J kg^{-1})
$I_{k' \to k}$ interphase momentum transfer from k' to k (kg m^{-2} s^{-2})
K_k^{eff} effective thermal diffusivity of the phase k (m^2 s^{-1})
K_g^l laminar gas thermal diffusivity (m^2 s^{-1})

K_g^t turbulent gas thermal diffusivity ($m^2\ s^{-1}$)
K_p^{kin} particle thermal diffusivity due to the particle agitation ($m^2\ s^{-1}$)
K_p^r particle thermal diffusivity due to the particle–particle radiative transfer ($m^2\ s^{-1}$)
k gas turbulent kinetic energy ($m^2\ s^{-2}$)
m_p particle mass (kg)
n_p particle number density (m^{-3})
Nu Nusselt number (–)
P_g gas pressure (Pa)
P_{ref} reference pressure (Pa)
P_α partial pressure of the species α (Pa)
Pr Prandtl number (–)
q_p^2 particle fluctuant kinetic energy ($m^2\ s^{-2}$)
q_{gp} fluid–particle velocity covariance ($m^2\ s^{-2}$)
Q_r interphase heat transfer rate (W)
R_g ideal gas constant ($J\ mol^{-1}\ K^{-1}$)
R_1, R_2, $R_{ch,1}$, $R_{ch,2}$ chemical reaction rates ($mol\ m^{-3}\ s^{-1}$)
$R_{mix,1}$, $R_{mix,2}$ turbulence mixing rates ($mol\ m^{-3}\ s^{-1}$)
$\mathbf{Re}_p$ particulate Reynolds number (–)
$R_{k,ij}$ kinetic stress tensor of phase k ($m^2\ s^{-2}$)
S_{rad_k} thermal radiation source term for the phase k ($W\ m^{-3}$)
S_{ext} external heat source (W)
S_p speedup (–)
T_k temperature of the phase k (K)
T_W constant wall temperature (K)
T_r relative temperature (K)
T_{ref} reference temperature (K)
T_{bulk} surrounding medium temperature (K)
$U_{k,i}$ velocity (i-component) of the phase k ($m\ s^{-1}$)
U_{mf} minimum fluidization velocity ($m\ s^{-1}$)
U_f gas fluidization velocity ($m\ s^{-1}$)
$U_{\sigma,i}$ interface (i-component) velocity ($m\ s^{-1}$)
$u_{k,i}$ instantaneous (i-component) velocity of the phase k ($m\ s^{-1}$)
$V_{d,i}$ fluid–particle (i-component) turbulence drift velocity ($m\ s^{-1}$)
$v_{r,i}$ instantaneous (i-component) relative velocity ($m\ s^{-1}$)
$V_{r,i}$ relative (i-component) velocity ($m\ s^{-1}$)
W_α species molar mass ($kg\ mol^{-1}$)
W_g gas mixture molar mass ($kg\ mol^{-1}$)
X_α species molar fraction (–)
$\mathbf{X}_d$ mean particle number per unit mass (kg^{-1})
Y_α species mass fraction (–)

GREEK LETTERS

α_k volume fraction of the phase k (–)
β_α radiative absorption coefficient of the species α ($m^{-1}\ atm^{-1}$)
β_p radiative absorption coefficient of the particulate phase (m^{-1})

γ_p particle disappearance rate (m^{-3} s^{-1})
$\boldsymbol{\Gamma}_k$ interphase mass transfer rate from k' toward k (kg m^{-3} s^{-1})
δ_{ij} Kronecker symbol (–)
ϵ_{bulk} surrounding medium radiation emissivity (–)
ϵ_p particle radiation emissivity (–)
ϵ_W wall radiation emissivity (–)
ε gas turbulent dissipation rate (m^2 s^{-3})
ε_p particle kinetic energy dissipation rate (m^2 s^{-3})
ε_{gp} fluid–particle covariance dissipation rate (m^2 s^{-3})
θ_r instantaneous relative temperature (K)
$\boldsymbol{\Theta}_{g,ij}$ viscous stress tensor (kg m^{-1} s^{-2})
$\boldsymbol{\Theta}_{p,ij}$ collisional stress tensor (kg m^{-1} s^{-2})
κ_p^{eff} particle turbulent effective diffusivity (m^2 s^{-1})
κ_p^{kin} particle turbulent kinetic diffusivity (m^2 s^{-1})
κ_p^{col} particle turbulent collisional diffusivity (m^2 s^{-1})
λ_k thermal conductivity of phase k (W m^{-1} K^{-1})
μ_k dynamic viscosity of the phase k (Pa s)
ν_k kinematic viscosity of the phase k (m^2 s^{-1})
ν_g^t turbulent gas viscosity (m^2 s^{-1})
ν_{gp}^t turbulent gas–particle viscosity (m^2 s^{-1})
ν_p^{kin} particle kinetic viscosity (m^2 s^{-1})
ν_p^{col} particle collisional viscosity (m^2 s^{-1})
$\Pi_{k'\to k}$ interphase enthalpy exchange from k' to k (J m^{-3} s^{-1})
Π_{q_p} interphase turbulent kinetic energy transfer rate (kg m^{-1} s^{-3})
$\Pi_{q_{gp}}$ interphase fluid–particle covariance interaction term (kg m^{-1} s^{-3})
ρ_k mass density of the phase k (kg m^{-3})
σ_{SB} Stefan–Boltzmann constant (W m^{-2} K^{-4})
σ_α^t turbulent species Schmidt number (–)
$\boldsymbol{\Sigma}_{k,ij}$ effective stress tensor of the phase k (kg m^{-1} s^{-2})
τ_g^t turbulence characteristic time (s)
τ_{gp}^T mean particle thermal relaxation time (s)
τ_{gp}^F mean particle dynamic relaxation time (s)
τ_p^c interparticle collision time (s)
τ_{gp}^t eddy-particle interaction time (s)
$\tau_{ch,1}$, $\tau_{ch,2}$ chemical reaction times (s)
τ_{CH_4} methane reaction time (s)
τ_{mix} time of turbulent mixing (s)
τ_{ref}, τ_n effective core CPU times (s)
$\boldsymbol{\Phi}$ air factor (–)
χ_{CH_4} methane conversion (–)
$\boldsymbol{\Psi}_\alpha$ mass change rate of the gaseous species α (kg m^{-3} s^{-1})
$\{\psi_k\}_k$ average of any k-phase property ψ_k (units of ψ_k)
$\langle\psi_k\rangle_k$ density-weighted average of any k-phase property (units of ψ_k)
ψ_k'' fluctuating component of the variable ψ_k (units of ψ_k)

1. INTRODUCTION

Multiphase reactive flows concern many industrial applications inherent to the energy conversion, waste recycling, polymerization, and many other manufacturing processes. In these flows, several phases (solid, gas, and/or liquid) coexist and interact with each other. Chemical reactions may then occur either between species of the same phase or between different phases' compounds. In all these processes, hydrodynamic interactions play an important role since they affect the interphase mass and heat transfers, especially in dense regimes (high particulate phase concentration). The complexity of such flows depends on the specificity of each phase (density, composition, size distribution, etc.), the nature of the reactions (homogeneous or heterogeneous) and the operating conditions (temperature, mass flow rate, etc.). Several different systems may be used in these industrial applications as, for example, boilers, furnaces, stirred tanks, fixed or fluidized beds. Fluidized bed reactors are widely employed because of their combustion efficiency and temperature control. They are used in many industrial processes, including conventional and alternative solid-fuels combustion (Anthony, 1995; Scala and Salatino, 2002), biomass gasification (Gómez-Barea and Leckner, 2010), sludge incineration (Ogada and Werther, 1996), food processes (Sotudeh-Gharebagh and Mostoufi, 2004), as well as in emerging industrial technologies for steam or power generation, as the Chemical Looping Combustion (Pröll et al., 2009), for example. For catalytic combustion applications, fluidized beds are also preferred to fixed beds due to the enhanced interphase heat transfer which limits catalyst overheating, so avoiding catalyst thermal deactivation (Foka et al., 1994; Iamarino et al., 2006). In fluidized beds, gaseous reactions may occur when the system is fed with gaseous fuel, or when volatiles are released from the solid fuel. Inert solid particles then ensure mixing and heat exchanges thus improving the reactor efficiency. Most of the aforementioned processes use natural gas as a fuel because it contains low levels of sulfur and produces lower NO_x emissions when burning at relatively low temperature (Foka et al., 1994). Extensive experimental and theoretical studies on the gaseous combustion in fluidized beds, often performed at the bubbling fluidization regime, have been reported in the literature. These studies revealed the crucial role played by the bed temperature on the characterization of the combustion location and the gas composition. Dounit et al. (2001, 2008) pointed out the gradually moving of the combustion zone toward the bed surface and inside the bed as the bed temperature increases. For premixed natural gas–air lean combustion, a critical temperature was experimentally observed between 800°C and 850°C below which, fuel mainly burns in the projection zone. They also pointed out a combustion zone displacement depending on the particle diameter and the gas flow rate. Experiments over a broad range of temperatures were performed and models proposed for the characterization of the reactor behavior (Dounit et al., 2001, 2008; Pre et al., 1998). The question of the existence of a critical temperature, which depends on the operating conditions, was discussed by several authors in the literature, for premixed or nonpremixed mixtures and different fuels (see, e.g., Sotudeh-Gharebagh et al., 2007; van der Vaart, 1988). The study conducted by Żukowski (2003) with natural gas, suggested two main critical temperatures, defining three temperature ranges, each one corresponding to a different regime. For bed temperatures lower than a first critical temperature (550°C) the fuel burns over the bed, while for bed temperatures higher than a second critical temperature (860°C) the combustion moves inside the bed, toward the distributor. Between the two critical temperatures, a more complex regime exists in which combustion takes place into the bed, close to the bed surface, because of the ignition of bubbles which are leaving the bed surface. This regime is accompanied by acoustic effects related to ignited bubble explosions (Żukowski, 2003). In general, acoustic magnitude and frequency depend on the bed temperature and on the fuel used for the combustion (Żukowski, 1999); different regimes also lead to different concentrations of NO compared to NO_2 (Baron et al., 2002; Żukowski, 2003). The recent work of Baron et al. (2012) confirms the evidence of the aforementioned regimes and points out a dependency of the ignition mode on the bubble residence time compared to the thermal ignition time. By experiments conducted in fluidized beds, at incipient conditions, using stoichiometric premixed air–methane and air–propane mixtures, Hesketh and Davidson (1991) concluded that, into the bed, inert particles inhibit the combustion because of the free radical destruction at their surface. By a consequence, they suggested that combustion can only take place in the bubble or at the bed surface when the bed temperature is lower than a critical one. Gaseous combustion in fluidized bed thus differs from the combustion in one-phase flows. Particles may indeed act as a quenching on the combustion, modifying the ignition delay, and the final temperature as well. This involves complex physical mechanisms related to the coupling between the bed hydrodynamic and the gaseous reactions, which are still poorly understood. These mechanisms are somewhat intricate to model and also difficult to reproduce in laboratory, especially when the apparatus size affect the results (Rüdisüli et al., 2012). Lab-scale experiments and prototype tests are, however, extremely important in order to provide information about the overall process. Theoretical and numerical models may then be conceived on the basis of the experimental observations and used to predict the reactive flow, at industrial scale as well. Among the different approaches available in the literature, the unsteady three-dimensional modeling additionally gives local and instantaneous information, making it possible to gain insight in the local and instantaneous reactive flow behavior, therefore allowing the computational fluid dynamics (CFD) to provide recommendation for design upgrade in the stage of scaling-up from pilot to industrial plants. The unsteady 3D CFD is now emerging due to the hardware performance increasing and thanks to the use of massive parallel computations. 3D numerical simulations of reactive fluidized beds are now feasible, in transient conditions as well (Alobaid et al., 2015; Hamidouche, 2017; Loha et al., 2014; Parker, 2014). In the present study, unsteady 3D numerical simulations of a premixed air–methane combustion in a dense fluidized bed combustor were performed. Numerical simulations were realized using an Euler–Euler approach and results assessed against experimental measurements (Dounit, 2001; Dounit et al., 2001, 2008). A further analysis at mesoscopic and macroscopic scales was then carried out in order to gain insight in the local behavior of the reactive flow and contribute to the understanding of the combustion features in such reactive dense fluidized beds.

2. BRIEF DESCRIPTION OF THE EXPERIMENTAL CAMPAIGN

The present study is a numerical investigation of air–methane combustion in a dense fluidized bed reactor for which experimental results are available by the works of Dounit (2001) and Dounit et al. (2001, 2008). In these works, experiments were performed at different operating points. In this section, they are briefly recalled. A sketch of the experimental setup is given in Fig. 1. It consists of a reactor of height 1400 mm and inner diameter 180 mm, supplemented by a disengagement section of diameter 360 mm. In order to maintain a constant temperature in the bed, cooling air is made to circulate in a double shell. The cooling air circuit has a pneumatic valve regulated by a PID system supplied of a thermocouple which is located in the bed. The experiments were conducted using chemically

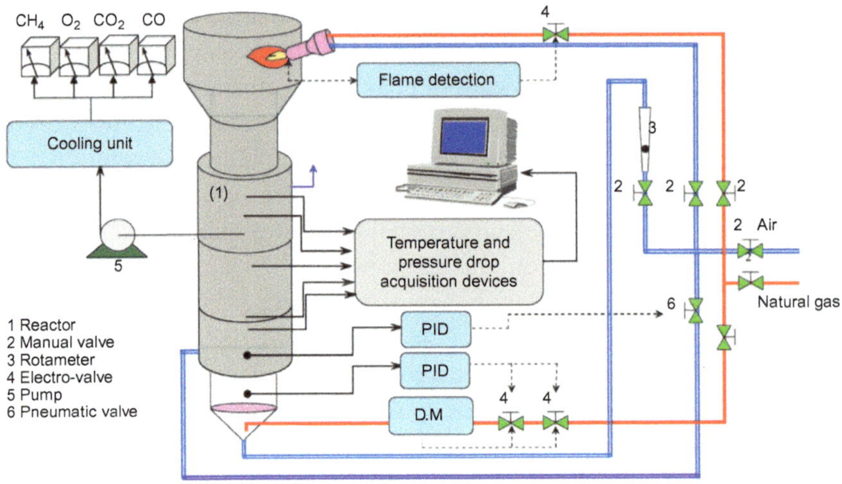

Fig. 1 Sketch of the experimental setup excerpted by Dounit et al. (2008).

inert sand particles and natural gas as a fuel, with 97% of methane content. The reactor was equipped with a wind box, located between a primary and a secondary distributors, which ensures premixing of air with natural gas at the reactor inlet. In addition, the reactor was axially fitted with temperature sensors and gas-sampling water-cooled tubes at different locations above the primary distributor. The sampling tubes were connected to the cooling unit in order to eliminate steam. The molar fractions of CH_4, CO_2, and CO were measured by means of infrared-type analyzers while for O_2 paramagnetic-type analyzers were used. A detailed description of the experiments may be found in the aforementioned publications (Dounit, 2001; Dounit et al., 2001, 2008). As summarized in Table 1, experiments were performed for several operating points defined by varying some relevant parameters such as the solid mass, the superficial gas velocity, the mean particle diameter, and the air factor; the latter is defined as the ratio between the volume of air supplied and that required for stoichiometric combustion of methane. In addition, in the experiments, the C1 operating point was investigated over several bed temperatures ranging between 650°C and 850°C. Further experimental quantities of interest are reported in Table 2.

In Section 5, the results obtained from the numerical simulations, carried out at the conditions corresponding to C1, will be presented. For this reason, only the experimental results related to such selected operating point will be reported in this work and used for comparison purposes.

Table 1 Operating Conditions of the Air–Natural Gas Combustion From the Experiments (Dounit, 2001; Dounit et al., 2008)

Test	Solid Mass (kg)	U_f/U_{mf} at 20°C	Air Factor Φ	Particle Diameter (μm)	Variable Parameter
C1	12	2	1.2	315–400	Reference test
C2	9	2	1.2	315–400	Solid mass
C3	15	2	1.2	315–400	
C4	12	3	1.2	315–400	Gas velocity
C5	12	4	1.2	315–400	
C6	12	2	1.0	315–400	Air factor
C7	12	2	1.5	315–400	
C8	12	2	1.2	80–125	Particle diameter
C9	12	2	1.2	500–630	

Table 2 Properties of Sand Particles and Their Fluidization (Dounit, 2001; Dounit et al., 2008)

Solid Particles	Units	
Particle density	kg m^{-3}	2650
Mass heat capacity	J kg K^{-1}	1050
Thermal conductivity	W kg K^{-1}	1.75
Radiative emissivity	—	0.60
Minimum fluidization velocity	m s^{-1}	
– for 80–125 μm-sized particles		0.015
– for 315–400 μm-sized particles		0.083
– for 500–630 μm-sized particles		0.260

3. MODELING DESCRIPTION

The modeling used for predicting the natural gas combustion in the fluidized bed reactor is here presented. It consists in the Eulerian approach used for simulating the hydro-thermodynamics of the reactor and the models/closures inherent to the reactive flow. A brief description of the overall modeling is given in the next sections. Further details are provided in Appendix.

3.1 Eulerian Approach

The interactions between the dispersed phase, composed of solid particles of equal constant diameter, and the gas mixture of N species are predicted in an Eulerian framework using the two-fluid formalism. According to this approach, two distinct sets of equations (one for each phase) are solved. These sets, including mass, momentum, and enthalpy transport equations, are coupled by specific closures modeling interphase mass, momentum, and energy transfers. Since particles do not react with gas, no mass transfer occurs and the related closures are set to zero in the balance equations. For each phase, the system of equations is written as follows:

$$\frac{\partial}{\partial t}(\alpha_k \rho_k) + \frac{\partial}{\partial x_j}(\alpha_k \rho_k U_{k,j}) = 0 \quad (1)$$

$$\alpha_k \rho_k \left(\frac{\partial U_{k,i}}{\partial t} + U_{k,j} \frac{\partial U_{k,i}}{\partial x_j} \right) = -\alpha_k \frac{\partial P_g}{\partial x_i} + \alpha_k \rho_k g_i + I_{k' \to k,i} + \frac{\partial \Sigma_{k,ij}}{\partial x_j} \quad (2)$$

$$\alpha_k \rho_k \left(\frac{\partial H_k}{\partial t} + U_{k,j} \frac{\partial H_k}{\partial x_j} \right) = S_{rad_k} + \Pi_{k' \to k} + \frac{\partial}{\partial x_j} \left(\alpha_k \rho_k K_k^{eff} \frac{\partial H_k}{\partial x_j} \right). \quad (3)$$

In the above equations, α_k represents the volume fraction of the phase k, which may be either the gas ($k = g$) or the particulate phase ($k = p$), ρ_k and $U_{k,j}$ are the phase density and velocity, respectively, and H_k is the phase mass enthalpy. $I_{k' \to k,i}$ and $\Pi_{k' \to k}$ account, respectively, for the interphase (gas–solid) momentum and energy transfers due to the local perturbations induced by the particles; in both these terms k' stands for the phase other than k, therefore denoting the effects of the phase k' on the phase k. $\Sigma_{k,ij}$ is the sum of the viscous and the turbulent stress tensors for the gas, or the particle kinetic and the collisional stress tensors for the solid. In Eq. (3), S_{rad_k} represents the heat exchange due to thermal radiation in the reactor. Its model is given in Section 3.3. The last term in Eq. (3) is written in the frame of a gradient approximation based on an effective thermal diffusivity, K_k^{eff}. The latter is composed of two contributions: for the gas they are the laminar and the turbulent thermal diffusivity ($K_g^{eff} = K_g^l + K_g^t$), for the particles they are the thermal diffusivity due to the particle agitation and that due to the radiative transfer within the solid phase, written in the frame of the Rosseland approximation ($K_p^{eff} = K_p^{kin} + K_p^r$). In the present work, the contribution K_p^r is neglected. Details about the approach and its closures are given in Appendixes A.1, A.2, and A.3. They may also be found in the works of Simonin (2000), Gobin et al. (2003), and Konan et al. (2010a,b).

An additional set of $N-1$ species mass fraction equations (4) and the conservation law (5) are used for predicting the gaseous species evolutions:

$$\alpha_g\rho_g\left(\frac{\partial Y_\alpha}{\partial t}+U_{g,j}\frac{\partial Y_\alpha}{\partial x_j}\right)=\frac{\partial}{\partial x_j}\left(\alpha_g\rho_g D_\alpha^{eff}\frac{\partial Y_\alpha}{\partial x_j}\right)+\Psi_\alpha, \tag{4}$$

$$\sum_{\alpha=1}^{N} Y_\alpha = 1. \tag{5}$$

The gas density is computed using the state law:

$$\rho_g=\frac{P_{ref}W_g}{R_g T_g}=\frac{P_{ref}}{R_g T_g}\left(\sum_{\alpha=1}^{N}\frac{Y_\alpha}{W_\alpha}\right)^{-1}. \tag{6}$$

In Eq. (4), $D_\alpha^{eff}=D_\alpha^l+D_\alpha^t$ is an effective diffusivity composed of a laminar and a turbulent contribution. The latter is defined as the ratio between the turbulent viscosity and a turbulent species Schmidt number which is taken equal to unity in this work: $D_\alpha^t=\nu_g^t/\sigma_\alpha^t$. The source term Ψ_α in Eq. (4) depends on the kinetics and reaction mechanisms retained for the reaction. Its modeling is detailed in the next section.

3.2 Methane–Air Premixed Combustion Modeling

The methane combustion is described by the following global two-step mechanism:

$$CH_4+3/2\ \ O_2\rightarrow CO+2H_2O \tag{7}$$

$$CO+1/2\ \ O_2\rightarrow CO_2. \tag{8}$$

The gaseous mixture is thus composed of $N=6$ species which are methane, oxygen, nitrogen, carbon dioxide, carbon monoxide and water vapor. Since the atomic species are always conserved during any chemical process, and neglecting the effect of the laminar diffusivity variation between the species, it was decided to transport the atomic mass fraction of C, H, and N, in order to reduce as much as possible the use of the source terms in the numerical solution. The atomic mass fractions are related to the mass fractions of the species composing the mixture by the following relations:

$$\begin{cases} Y_C = 12/16 \quad Y_{CH_4} + 12/28 \quad Y_{CO} + 12/44 \quad Y_{CO_2} \\ Y_H = 4/16 \quad Y_{CH_4} + 2/18 \quad Y_{H_2O} \\ Y_N = Y_{N_2} \end{cases} \tag{9}$$

Additional transport equations for CH_4 and CO are then used. The carbon dioxide and water vapor mass fractions (Y_{CO_2} and Y_{H_2O}) are deduced from the transported species, thanks to the system of Eq. (9) and the oxygen mass fraction is eventually computed according to the conservation law $\sum_k Y_k = 1$. The source terms for the atomic species mass fractions are therefore set equal to zero ($\Psi_C = \Psi_H = \Psi_N = 0$), whereas that of CH_4 and CO are modeled from the selected kinetics for the two-step mechanism (7) and (8). In this work, the following two-step reaction kinetic proposed by Dryer and Glassman (1973) was retained:

$$R_1 = 10^{13.2} \exp\left(-\frac{48,400 \times 4.18}{R_g T_g}\right) [CH_4]^{0.7} [O_2]^{0.8} \times 10^{-3}, \tag{10}$$

$$R_2 = 10^{10.75} \exp\left(-\frac{43,000 \times 4.18}{R_g T_g}\right) [CO][H_2O]^{0.5} [O_2]^{0.25} \times 10^{-4.5}. \tag{11}$$

R_1 and R_2 are the kinetic rates (in mol/m^3/s) and [–] the species molar concentrations (mol/m^3). The vanishing rate of methane (in kg/m^3/s) due to the reaction (7), is modeled according to Eq. (10) as

$$\Psi_{CH_4} = -\alpha_g W_{CH_4} R_1, \tag{12}$$

and the rate of change of carbon monoxide, which accounts for both appearance and vanishing of such a species due to reactions (7) and (8), is modeled as

$$\Psi_{CO} = \alpha_g W_{CO} (R_1 - R_2). \tag{13}$$

In order to assess numerical results against experimental measurements, and for an analysis purpose as well, the methane conversion is defined and computed as follows:

$$\chi_{CH_4} = 1 - X_{CH_4} / (X_{CH_4} + X_{CO_2} + X_{CO}), \tag{14}$$

where X_α is the molar fraction of the species α, obtained according to the expression

$$X_\alpha = Y_\alpha W_g / W_\alpha. \tag{15}$$

3.3 Thermal Radiation Model

A simple thermal radiation model was used in this study. Such a model assumes that each radiating gas contained in a given volume of control has an unobstructed isotropic view of the surrounding cold medium (Barlow et al., 2001), wall or solid particles in the present study. Then, the radiation heat loss rate per unit volume may be written as follows:

$$S_{rad,g} = 4\sigma_{SB}\sum_{\alpha=1}^{n} P_\alpha \beta_\alpha \epsilon_{bulk}\left(T_{bulk}^4 - T_g^4\right), \tag{16}$$

where σ_{SB} is the Stefan–Boltzmann constant (5.669×10^{-8} W m^{-2} K^{-4}). P_α is the partial pressure (in atmosphere units) and β_α the Planck mean absorption coefficient (m^{-1} atm^{-1}) of the species α. T_{bulk} and ϵ_{bulk} are the surrounding medium temperature and emissivity, respectively. T_{bulk} is either the local mean particle temperature T_p or the reactor wall temperature T_W, according to a critical value of the solid volume fraction. In this study, the following condition was imposed:

$$\left(T_{bulk}, \epsilon_{bulk}\right) = \begin{cases} \left(T_W, \epsilon_W\right) & \text{if } \alpha_p \leq 0.02, \\ \left(T_p, \epsilon_p\right) & \text{otherwise}, \end{cases} \tag{17}$$

where ϵ_p and ϵ_W are the particle and the wall emissivities, respectively. Four species ($n = 4$ in Eq. 16) were retained in the calculation and they are water vapor, carbon dioxide, carbon monoxide, and methane. Their absorption coefficients are made varying as a function of the gas temperature according to curve fits generated for temperatures between 300 and 2500 K (Barlow et al., 2001; Grosshandler, 1993):

$$\begin{aligned}
\beta_{CH_4} &= C_{0,CH_4} + C_{1,CH_4} T_g + C_{2,CH_4} T_g^{\,2} + C_{3,CH_4} T_g^{\,3} + C_{4,CH_4} T_g^{\,4}, \\
\beta_{CO} &= C_{0,CO} + C_{1,CO} T_g + C_{2,CO} T_g^{\,2} + C_{3,CO} T_g^{\,3} + C_{4,CO} T_g^{\,4}, \\
\beta_\alpha &= C_{0,\alpha} + C_{1,\alpha}\left(1000/T_g\right) + C_{2,\alpha}\left(1000/T_g\right)^2 + C_{3,\alpha}\left(1000/T_g\right)^3 \\
&\quad + C_{4,\alpha}\left(1000/T_g\right)^4 + C_{5,\alpha}\left(1000/T_g\right)^5; \\
\alpha &= \{CO_2, H_2O\}.
\end{aligned} \tag{18}$$

using the constant values as given in Tables 3 and 4. Concerning the solid phase, the heat exchange due to the particles' radiation with (primarily) the

Table 3 Values of the Constants Used for CH_4, CO_2, and H_2O Radiation Calculation (Barlow et al., 2001)

Species	CH_4	CO_2	H_2O
	Gas Temperature Range, $T_g \in [300, 2500]$ in K		
C_0	+6.6334000	+18.7410	−0.23093
C_1	−0.0035686	−121.310	−1.12390
C_2	$+1.66820 \times 10^{-8}$	+273.500	+9.41530
C_3	$+2.5611 \times 10^{-10}$	−194.050	−2.99880
C_4	-2.6558×10^{-14}	+56.3100	+0.51382
C_5	(–)	−5.81690	-1.86840×10^{-5}

Table 4 Values of the Constants Used for CO Radiation Calculation (Barlow et al., 2001)

	CO	
Species	$T_g \leq 750$ K	$T_g > 750$ K
C_0	+4.78690	+10.0900
C_1	−0.06953	−0.01183
C_2	$+2.95775 \times 10^{-4}$	$+4.77530 \times 10^{-6}$
C_3	-4.25732×10^{-7}	-5.87209×10^{-10}
C_4	$+2.02894 \times 10^{-10}$	-2.53340×10^{-14}

reactor walls, in dilute regions, and with the gas, in dense regions, is taken into account by the following model:

$$S_{rad,p} = \begin{cases} 4\sigma_{SB}\beta_p\epsilon_W\left(T_W^4 - T_p^4\right) & \text{if } \alpha_p \leq 0.02, \\ -S_{rad,g} & \text{otherwise,} \end{cases} \tag{19}$$

with the particle absorption coefficient given by

$$\beta_p = \frac{3}{2}\frac{\epsilon_p \alpha_p}{d_p}. \tag{20}$$

3.4 Local Phase Temperature Determination

The mass enthalpy, H_g, of the gaseous mixture composed of N chemical species α, each one associated to a mass enthalpy H_α, may be related to the gas temperature T_g by the following expression:

$$H_g = \sum_{\alpha=1}^{N} Y_\alpha H_\alpha(T_g). \tag{21}$$

According to the Shomate equations, the mass enthalpy of each species can be expressed as a function of the gas temperature by the following polynomial

$$H_\alpha(T_g) - H_\alpha(T_{ref}) = \frac{1000}{W_\alpha}[A\,t_g + B\,t_g{}^2/2 + C\,t_g{}^3/3 + D\,t_g{}^4/4 - E/t_g + F - H], \tag{22}$$

in which $t_g = T_g/1000$ and A to H are coefficients for each species, taken from the National Institute of Standards and Technology (NIST) (Linstrom and Mallard, 2001). A comparison between the enthalpy as obtained from the solution of Eq. (3) for the gas phase and the enthalpy as obtained from Eqs. (21) and (22) makes it possible to find the corresponding gas temperature, T_g, using an iterative numerical method. Once the temperature is obtained, the mass heat capacity, C_{p_α}, of each species is updated thanks to the corresponding Shomate equation. The mass heat capacity of the mixture, C_{p_g}, is then obtained as $C_{p_g} = \sum_{\alpha=1}^{N} Y_\alpha C_{p_\alpha}(T_g)$. For the particulate phase, it was assumed that the mass heat capacity C_{p_p} does not depend on the particulate temperature T_p (its constant value is given in Table 2). Therefore, the particulate phase temperature is simply obtained by the relation

$$T_p = T_{ref} + H_p/C_{p_p}. \tag{23}$$

4. NUMERICAL SIMULATIONS

The unsteady three-dimensional numerical simulations of the fluidized bed reactor were performed using the Eulerian N-fluid modeling approach for fluid–particle turbulent polydispersed reactive flows implemented in NEPTUNE_CFD V1.08@Tlse version by IMFT (Institut de Mécanique des Fluides de Toulouse). NEPTUNE_CFD is a computational multiphase flow software developed in the framework of the NEPTUNE project, financially supported by CEA (Commissariat à l'Énergie Atomique), EDF (Electricité de France), IRSN (Institut de Radioprotection et de Sûreté Nucléaire), and AREVA-NP.

4.1 Code Description and Performances

The main numerical characteristics of NEPTUNE_CFD V1.08 are unstructured meshes with all types of cell, nonconforming connections, cell-center type finite volume method, calculation of co-localized gradients with reconstruction methods and distributed-memory parallelism by domain decomposition (MPI parallelization). The numerical solver is written in Fortran and C/C++ languages. The partial differential equations are discretized with a second-order centered scheme and the solution is time-advanced by a first-order scheme. The algorithm is based on an elliptic semiimplicit fractional step method using iterative linear solvers or direct matrix inversion (Méchitoua et al., 2003). The algorithm accounts for density variation as a function of pressure, enthalpy, and species mass fraction.

Numerical simulation of fluidized beds may be very expensive (mesh and CPU time), both at industrial and laboratory scales, in particular because of the presence of 3D unsteady structures of small size (clusters or bubbles) which need to be resolved. Simulations may need from thousands to millions of computational hours on supercomputer and consequently they require the use of High Performance Computing (HPC). In a previous study, NEPTUNE_CFD scalability was checked up to 1024 cores (Neau et al., 2010). The parallel performances of the code NEPTUNE_CFD, evaluated on a 3D uniform granular shear flow and on an industrial-scale reactor (a gas–solid fluidized bed), are here presented. Numerical simulations were carried out on a cluster in production: SC1 (SGI Altix ICE8200-EX based on Intel Xeon X5560 Quad Core processors) and SC2 (Bullx S6010 based on Intel Xeon X7560 Eight Core processors). The uniform shear flow consists of a 3D isothermal granular shear flow of a mono-disperse and a bi-disperse solid particle mixture (shear rate $= 20\ s^{-1}$). The computational domain is a cubic box of length $H_{cube} = 0.01$ m. The solid-to-gas density ratio is 1000. For the mono-disperse case, a particle diameter of 350 μm was chosen. For the bi-disperse case, a coarse-to-fine ratio between particle diameters of 8.75, typically encountered in industrial fluidized bed of catalytic polymerization, was retained. A particle–particle restitution coefficient of 0.8 was imposed. The gas phase was set as laminar and the particle agitation was accounted for using separate equations of the fluctuant particle kinetic energy for both the solid species (Fede et al., 2011). Gravity was not taken into account. The schematic of the system is shown in Fig. 2.

The evaluation of NEPTUNE_CFD parallel performances was carried out using the mesh referred to as 38M (a grid composed of 37,933,056 uniform-size hexahedra) from 128 to 4096 cores (1–32 nodes) on SC2. Three boundary conditions were selected: (i) right-side and left-side, inlet–outlet conditions for particles and gas with imposed velocity $U_x = 20(y - H_{cube}/2)$, (ii) front-side and back-side, symmetry condition for

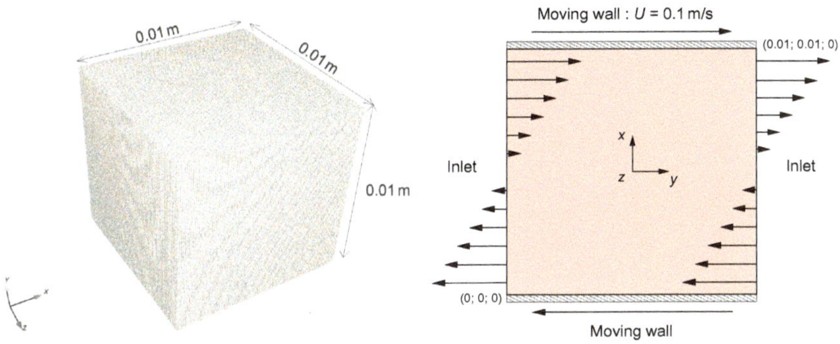

Fig. 2 Sketch and mesh of a 3D granular uniform shear flow.

particle and gas, and (iii) top-side and bottom-side, moving wall: $U_{wall} = 20(y_{wall} - H_{cube}/2)$. Moving wall was reproduced by a no-slip wall boundary (gas and particle): $U_{g,x} = U_{p,x} = U_{wall}$ and $\partial q_p^2/\partial n = 0$ (where q_p^2 is the fluctuant particle kinetic energy). An adaptative time step was used (computed from Courant and Fourier criteria). The following iterative solvers were selected: a Jacobi method for the velocity, a conjugated gradient for the pressure, and a biconjugate gradient stabilized for the volume fraction. After a transient phase, numerical simulations converged to a stationary state. Then, the numerical results were compared with the analytical 0D solution. The accuracy of NEPTUNE_CFD numerical results was found excellent and independent of parallelization, core number, and mesh size (here not shown). The speedup S_p was defined as the ratio between the elapsed time to execute a program on one node (ref=1node=8cores on SC1 and ref=1node=128cores on SC2) and that to execute the program on a set of concurrent n nodes ($n \times$ ref), and the efficiency E_f was defined as the ratio between the speedup and n: $S_p = \tau_{ref}/\tau_n = \tau_{8cores}/\tau_n$ and $E_f = S_p/n$ where τ_{ref} is the effective core CPU time of the parallel job on one node (ref) and τ_n is the effective core CPU time of the parallel job on n nodes ($n \times$ ref). The restitution time of a simulation is divided into computation time, communication, and waiting time (send/receive/barrier MPI) and Input/Output time (zero in this case). Basically, increasing the core number corresponds to decreasing the cell number per core. However, a decrease of computation time per core leads to an increase of communication time between cores. The core limit number is reached when the communication time increases quicker than the computation time decreases. Thus, there is an optimal number of cores for a given mesh and a given number of partial differential equations. Fig. 3 shows the speedup and the efficiency, obtained

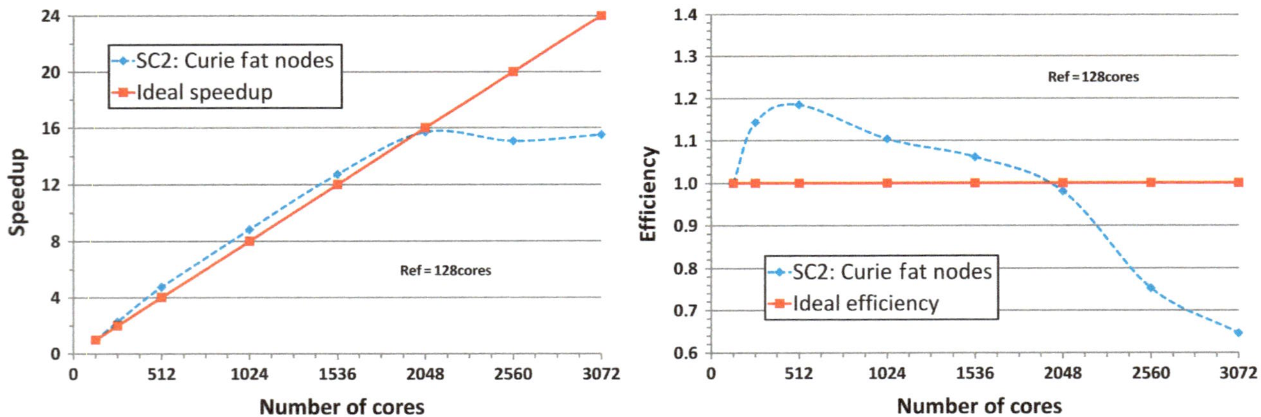

Fig. 3 Parallel performances assessed using 38M mesh on the mono-disperse case.

for mesh 38M, defined with ref=128cores. The solid line represents the ideal speedup and efficiency. An excellent HPC efficiency even for large core number (2,048 cores, 16 nodes) is observed. The speedup is super-linear up to 1536 cores (such a behavior comes from cache effect and memory bandwidth) and linear for 2048 cores. Efficiency is also greater or equal to unity. This study confirms the very good parallel performances of NEPTUNE_CFD. Using mesh 38M, performance degradation begins when the cell number per core is lower than 18,500. Previous studies with MPINSIDE (Neau et al., 2010) showed that MPI communications (barrier before MPI Allreduce), due to iterative solvers, increase quickly with the core number. NEPTUNE_CFD strong scaling is excellent up to 2048 cores on SC2, using a minimum cell number per core (18,500 cells per core). Using a cluster with "thin" nodes and an efficient MPI, a scalability up to 3584 cores is obtained (not shown). Fig. 4 shows NEPTUNE_CFD speedup (ref=128cores) for the bi-disperse test case. NEPTUNE_CFD performances are excellent up to 2560 cores (that is, while the cell number per core is greater than 14,800). For this bi-disperse numerical simulation, NEPTUNE_CFD solve 16 partial differential equations instead of 10 accounted for in the previous case. Therefore, computation versus communication balance improves and the minimum number of cells per core decreases. This study on a uniform granular (mono/bi-disperse) shear flow shows both the accuracy and the excellent scalability of NEPTUNE_CFD for large-enough problem sizes.

The second test case is a 3D unsteady dense fluidized bed at industrial scale. The reactor (about 30 m high and 5 m wide) was meshed using

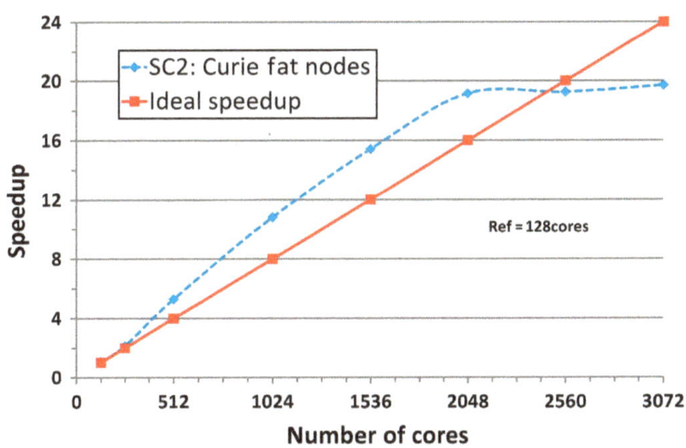

Fig. 4 Parallel performances assessed using 38M mesh on the bi-disperse case.

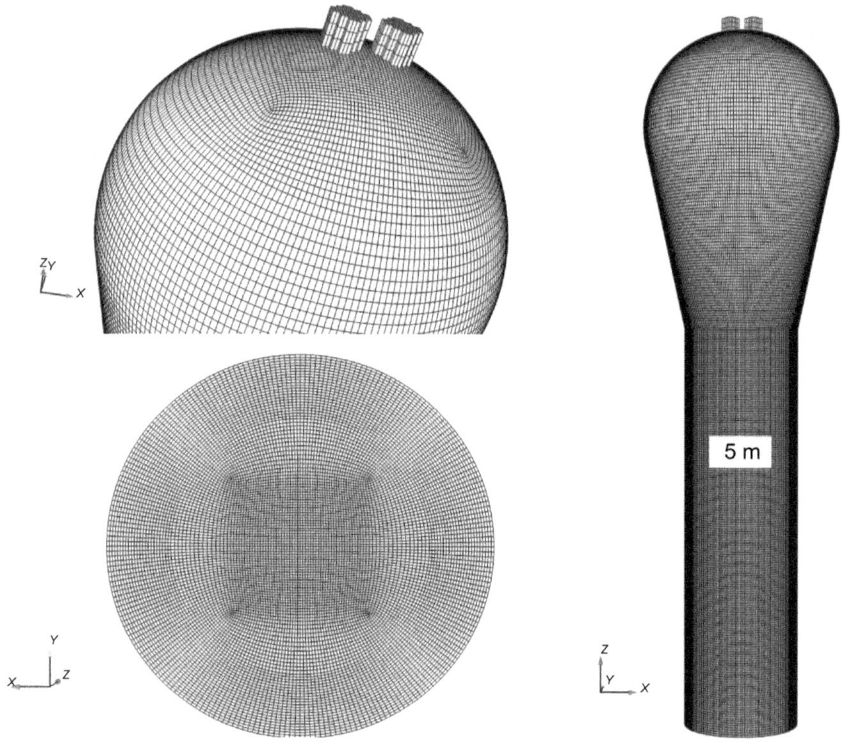

Fig. 5 Mesh of the industrial fluidized bed.

Table 5 Powder and Gas Properties

Phase	Gas	Solid
Density (kg/m^3)	21	850
Viscosity × 10^{-5} (Pa×s)	1.54	–
Pressure (bar)	20	–
Median diameter (μm)	–	1600
Solid mass (kg)	–	80,000
Restitution coefficient (–)	–	0.9

3,150,716 hexaedra of approximate size $\Delta x = \Delta y \sim 30$ mm and $\Delta z =$ 90 mm (see Fig. 5). Powder/gas properties and operating point are given in Table 5. A $k - \varepsilon$ model, extended to fluid–particle flows, was used for the gas (Appendix A.1). For the solid phase, a coupled transport equation system was solved for the fluctuant particle kinetic energy and the fluctuant fluid–particle velocity covariance q_{gp} (Appendix A.3). At the bottom, an

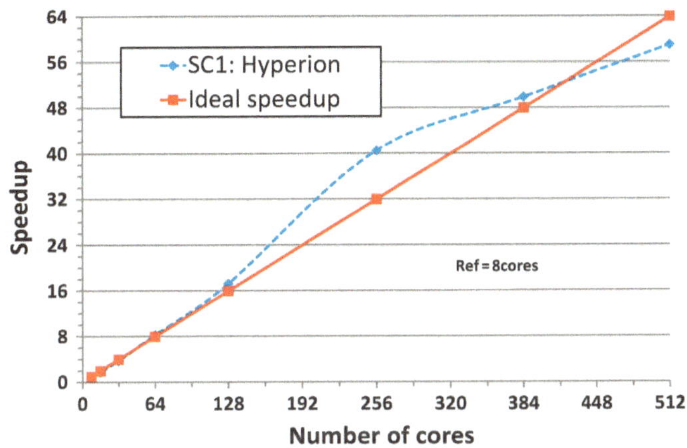

Fig. 6 Parallel performances assessed on the industrial reactor.

inlet boundary condition, corresponding to a fluidization velocity $U_f = 0.67$ m/s, was imposed for the gas. Free outlets conditions were chosen, at the top, for both the phases. At the wall, a friction condition was imposed for the gas and a slip condition was selected for the solid. NEPTUNE_CFD HPC was evaluated on a restart simulation of 500 iterations after fluidized bed destabilization. Simulations were carried out on the cluster SC1. The code speedup is given in Fig. 6. The restitution time decreases up to 512 cores. For this simulation, solving 12 equations, NEPTUNE_CFD HPC performances are found excellent up to 384 cores, i.e., while the cell number per core is higher than 8200. This simulation of a fluidized bed reactor illustrates perfect NEPTUNE_CFD scalability at industrial scale as well.

Considering such a 3D reactor, setting a maximum CPU time per core of 15 days, one can evaluate the maximum number of mesh cells which can be used to perform a numerical simulation of physical time 20 s. Fig. 7 shows the past evolution and an estimated future evolution of this maximum number of mesh cells. Between 1995 and 2008 the evolution was due to hardware performance increasing (processor, RAM). Since 2008, parallelization by decomposition domain allowed the use of larger meshes. The current step is the massively parallel computation (thousands of cores) with multigrid solvers and many cores processors. The 3D gas–solid fluidized bed reactor used for assessing the code scalability on a realistic test case, were further numerically simulated using a mesh of 108,000,756 hexahedra of approximate size $\Delta x \sim \Delta y \sim \Delta z \sim 1$ cm. Numerical simulations were performed on a supercomputer using a cluster BULLx DLC (12,240 cores—612 nodes) based on Intel(r) IVYBRIDGE 2.8 GHz processors. The integrality of such a supercomputer was used during 72 h. To obtain 20 s of physical time,

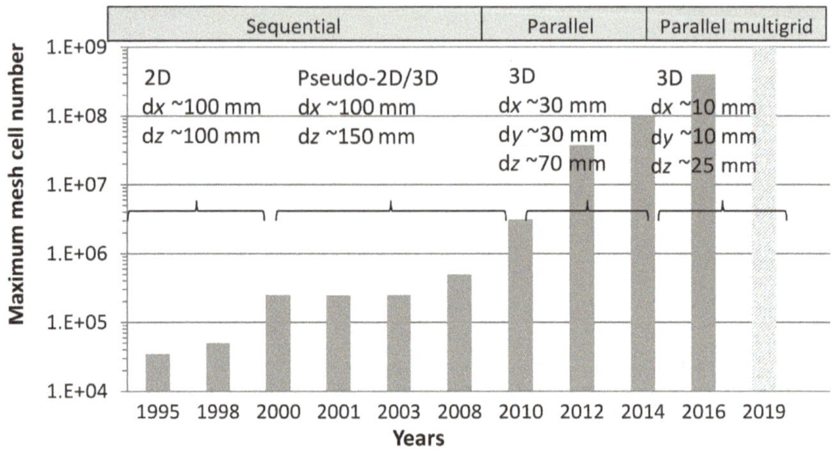

Fig. 7 History and forecast of the code HPC capabilities for fluidized bed reactors.

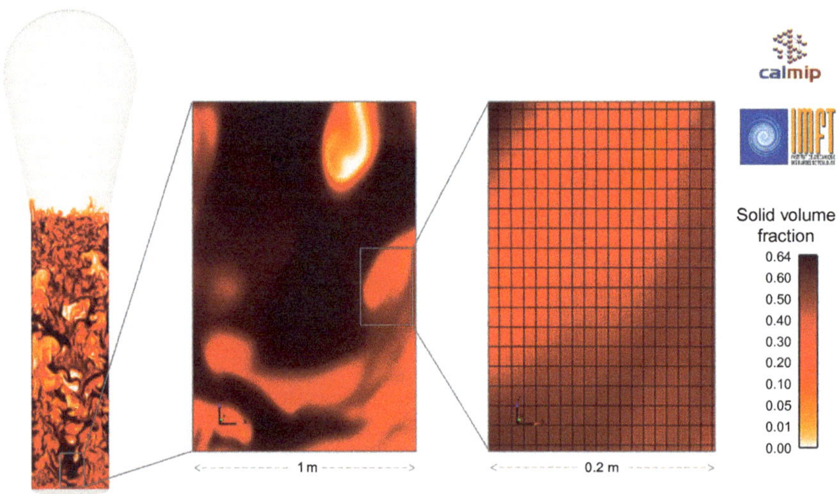

Fig. 8 Fluidized bed reactor computation using 12,000 cores (108,000,756 cells).

1 million CPU hours were spent using 12,000 cores (600 nodes), i.e., 9000 mesh cells per core. This case (Fig. 8) represents a feasibility demonstration of gas–solid computations with meshes of more than 100 millions of cells, using more than 10,000 cores. NEPTUNE_CFD parallel performances were found excellent on massively parallel computation.

4.2 Reactive Fluidized Bed Simulations

A sketch of the numerical configuration is shown in Fig. 9A. It consists of a cylindrical reactor of diameter 180 mm and height 1400 mm completed

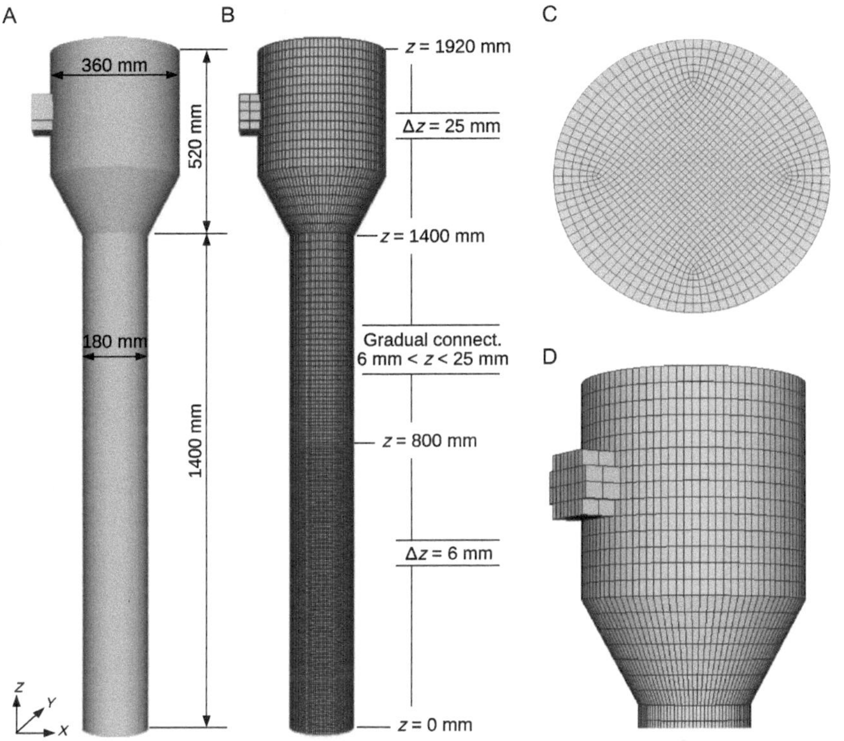

Fig. 9 Sketch and mesh of the fluidized bed reactor. (A) Reactor geometry, (B) full reactor mesh, (C) bottom inlet, and (D) top of the reactor showing the smokestack for exhaust gas.

Table 6 Mesh Characteristics: Approximated Cell Size and Number of Hexaedra

	Units in cm			
	Δx	Δy	Δz	No. of Cells
Mesh A	1.4	1.4	1.4	24,224
Mesh B	0.9	0.9	0.9	65,904
Mesh C	0.6	0.6	0.6	227,016
Mesh D	0.3	0.3	0.3	1,536,096

with an above disengaging section of diameter 360 mm and height 520 mm. The entire domain was meshed using a 3D grid based on four different refinements. The four meshes and their characteristics are summarized in Table 6. For all the meshes, Δz was set to the value indicated in Table 6 for $z \leq 80$ cm, and set as 2.5 cm for $z \geq 140$ cm; a gradual connection inside the range 80 cm $< z <$ 140 cm was then applied. As an example, Mesh C is shown in Fig. 9B–D.

All the simulations were run at the operating condition corresponding to C1 in the experiments (Table 1) at the selected bed temperature $T_{bed} = 700°C$. For all the simulations, conductive heat exchanges at the reactor wall, when taken into account, were considered up to $z = 0.8$ m, under the assumption of a constant wall temperature. However, two simulations were performed without conductive heat exchanges at the wall, over two different grids, in order to investigate its effect on the predictions. An additional simulation, using a finite rate chemistry/eddy dissipation approach (Xue and Fox, 2014), was also carried out, with the goal to assess the effectiveness of a combustion model to be used when coarse grids are employed. A summary of the simulation test cases is reported in Table 7.

In the present work, only one class of sand particles was retained, of diameter corresponding to the mean diameter of the sample used in the C1 experimental operating point (Table 1). The particle properties are reported in Table 8. In the experiments, a distributor composed of a perforated plate of 0.4% porosity is used. Its effect in the numerical simulations is reproduced by imposing a uniform gas flow rate, assuming a perfect air–methane mixing at the reactor inlet. Such an assumption is justified when the pressure drop induced by the distributor is comparable with the

Table 7 Simulation Test Cases

Case Designation	Mesh	Initial T_{bed} (°C)	Heat Exchange at Wall	T_W (°C)	Combustion Model
MESHA0	Mesh A	700	$\varphi_{k \to w} = 0$	(–)	DG
MESHB0	Mesh B				
MESHA1	Mesh A	700	(–)	707	DG
MESHB1	Mesh B				
MESHC1	Mesh C				
MESHD1	Mesh D				
MESHA2	Mesh A	700	(–)	707	min(DG,ED)

• $\varphi_{k \to w}$ is the imposed conductive heat exchange rate between the phase k and the reactor wall;

• In all the simulations, $\varphi_{k \to w} = 0$ for $z > 0.8$ m;

• DG refers to Dryer and Glassman mechanism;

• ED refers to the eddy dissipation approach;

• min(DG,ED) is the minimum between the DG kinetic rate and the mixing rate of the ED approach (see Section 5.5).

Table 8 Particle Properties
Particle Properties

Density, ρ_p	2650 kg m^{-3}
Mean diameter, d_p	350 µm
Particle–particle restitution coefficient, e_c	0.9
Radiative emissivity, ϵ_p	0.6

total drop pressure across the fluidized bed. No-slip boundary conditions for the mean particle velocity combined with zero-flux boundary conditions for the particle kinetic energy were imposed in the numerical simulations. These boundary conditions were found to be very effective in dense fluidized beds, reproducing the effects of spherical particles bouncing on a very rough wall (Fede et al., 2016). Finally, a friction boundary condition was imposed for the gas at wall, since a $k - \varepsilon$ model was used in this study. However, the latter should not play an important role on the predictions, as the turbulent contribution, in the bed, was found quite negligible in the numerical simulations.

At the reactor inlet, a lean mixture of methane and air, with an excess air factor of 1.2, was injected. Assuming that air is composed of 23.2wt% of oxygen and 76.8wt% of nitrogen, the corresponding inlet conditions for the species mass fractions are then deduced and the atomic mass fractions computed using Eq. (9). The inlet conditions are summarized in Table 9. Initially, for all the numerical simulations, both the gas and the particulate phases were set at the same temperature ($T = 700°C$). The reactor was filled with nitrogen and loaded with 12 kg of solid, corresponding to an initial particle volume fraction of 0.55.

5. RESULTS AND DISCUSSIONS

In this section, the results obtained from the unsteady 3D numerical simulations are presented and discussed with the goal to gain insight into the natural gas combustion occurring in the presence of an inert solid phase, in a dense regime. In order to assess the overall modeling, averaged quantities are computed and compared with available experimental measurements. Statistics are then computed at macroscopic and mesoscopic scales using, respectively, time-averaged and local-instantaneous fields. Such statistics are finally analyzed in order to gain insight into the gas–particle reactive flow.

Table 9 Gas Properties at the Reactor Inlet

Property at the Reactor Inlet	**Value**
Total flow rate	14.6 $Nm^3\ h^{-1}$
Air factor, Φ	1.2
Fluidizing velocity, U_f	$2U_{mf}$
Minimum fluidization velocity, U_{mf}	0.08 $m\ s^{-1}$
Temperature	298 K
Molecular species mass fractions	
CH_4	0.0461
O_2	0.2220
N_2	0.7319
H_2O, CO_2, and CO	0
Atomic species mass fractions	
C	0.034575
H	0.011525
N	0.731900
O	0.222000

All the simulations were run for 40 s, in order to reach a permanent regime, and then continued for 40 s in order to compute statistics. Indeed, when a permanent regime is reached, particles and fluid behave in a statistically stationary way and the time average of any quantity ξ_k may be computed as follows:

$$\overline{\xi_k} = \frac{\int_{\Delta t} \alpha_k \rho_k \xi_k \mathrm{d}t}{\int_{\Delta t} \alpha_k \rho_k \mathrm{d}t}, \tag{24}$$

with $\Delta t = 40$ s. Time-averaged gas and solid volume fractions or mass fluxes are obtained without weighting by $\alpha_k \rho_k$; however, for the sake of easy, the same notation will be retained. Spatial averages are also computed in selected zones of the reactor (in a selected volume or over a plane). They are denoted by the following operator $\langle \cdot \rangle$.

5.1 Preliminary Results

The results obtained for simulations MESHA0 and MESHB0 are first analyzed. Fig. 10 shows the time-averaged pressure drop along the reactor height (at the wall, for $x = 0.09$ m) as predicted from the numerical simulations, compared with the experimental data. Globally, the pressure drop slop is very well reproduced. Some differences in value are instead observed (and estimated to be around 5%) at the bottom of the bed. These differences may be attributed to the elutriation phenomenon and/or to the uncertainty in the experimental measurements. Numerical against experimental results of the time-averaged gas temperature in the reactor are shown in Fig. 11A for MESHA0 and in Fig. 11B for MESHB0. The vertical profiles are plotted at different x coordinates, from the center to the wall, and $y = 0$. In the experiments, the temperature is measured by means of temperature sensors located at several reactor heights, at the center of the reactor ($x = 0$, $y = 0$ in the numerical simulation).

The gas temperature represents one of the most relevant parameters in the fluidized bed combustor and needs to be carefully investigated. In their studies, Pre et al. (1998), Dounit (2001), and Dounit et al. (2001, 2008) showed that the combustion zone, denoted as the zone corresponding to a maximum in the mean gas temperature profiles, moves in the reactor according to the temperature of the dense bed. In particular, they showed that for bed temperatures higher than 800°C, the combustion mainly takes place into the bed, while for bed temperatures lower than 800°C, the combustion

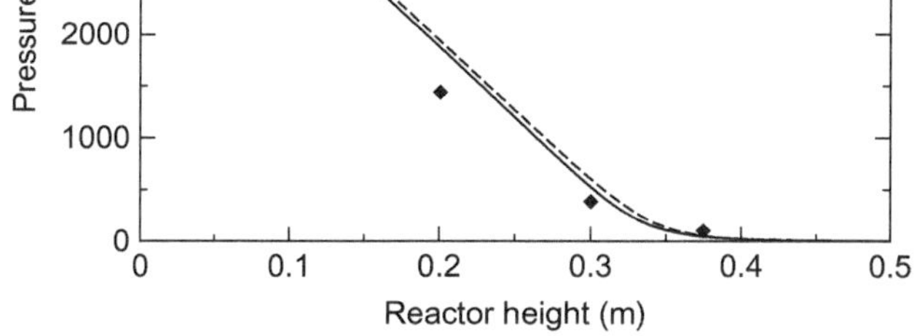

Fig. 10 Numerical, time-averaged, against experimental pressure drop at the wall ($x = 0.09$ m).

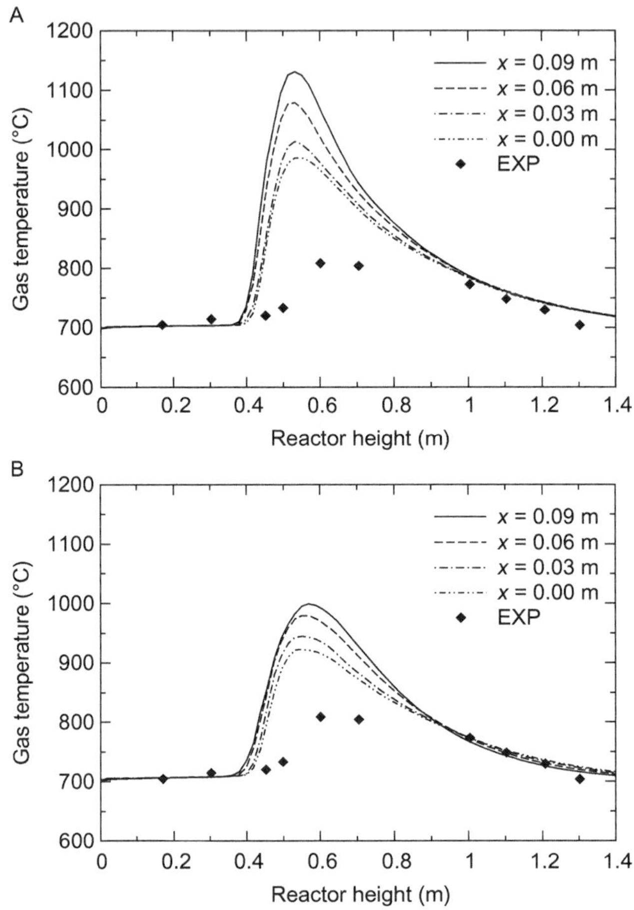

Fig. 11 Vertical profiles of the numerical time-averaged gas temperature against the experimental measurements. Results from simulations (A) MESHA0 and (B) MESHB0.

zone moves toward the bed surface and over. A critical temperature was pointed out in the range 800°C–850°C. These results agree with the results reported in the literature (see, e.g., Żukowski, 2003). At the operating point of 700°C, most of the combustion is expected to take place in the freeboard region. This is confirmed by the results illustrated in Fig. 11A and B. It is noteworthy, however, that the numerical simulations overestimate the gas temperature outside the bed, for $z > 0.4$ m, while its predictions are quite accurate into the bed. Furthermore, in both the simulations, radial inhomogeneities are observed above the bed surface. Such inhomogeneities are as much pronounced as the grid is coarse. Moreover, comparison between numerical results (Fig. 11A vs Fig. 11B) shows that the highest deviations from the experiments are obtained for the coarsest grid (Mesh A). Such a poor resolution has in fact an important effect on the modeling of the particulate phase. The correct prediction of the local solid amount is indeed crucial in order to accurately reproduce the combustion process. Using a more refined grid (Mesh B) makes it possible to improve the results. A difference of about 60°C at the center and 130°C near the wall is found when comparing Mesh B with Mesh A. Despite the ability of both the meshes to reproduce the good bed expansion for this class of particles (Geldart B), heat transfers are not accurately reproduced. Another difference between the numerical simulations and the experimental measurements concerns the location of the combustion zone. Indeed, in the numerical simulations, the combustion zone is slightly shifted toward the bed surface with respect to the experimental observations. This is deduced from the position (axial coordinate) corresponding to the gas temperature maximum (Fig. 11A and B). This point will be further discussed in Section 5.3. For both the simulations, comparisons between numerical and experimental species molar fractions are displayed in Fig. 12A. In the experiments, gas samples are taken at the center of the reactor. Therefore, the profiles of the time-averaged molar fraction are extracted at the reactor center as well. From the results, one can observe that oxygen and carbon dioxide, as predicted by the numerical simulations, do not match the experimental values at the top of the reactor. This is due to the fact that, in the numerical simulations, pure methane was taken into account while, in the experiments, natural gas with 97wt% methane content was used. The presence of small quantities of heavy hydrocarbons, such as ethane, propane, and butane, explains such a difference observed in the molar fractions of the reaction products. Furthermore, results show that the methane molar fraction is quite well reproduced into the bed while it is underestimated far from its surface. These results are consistent with the observations about the gas–temperature overestimate for $z > 0.4$ m. In this region, both the simulations predict faster fuel conversion, χ_{CH_4} (Eq. 14), compared to the experiments, as shown in Fig. 12B. With respect to the experiments, steeper curves of the methane conversion are indeed obtained by the simulations for $z > 0.4$ m. However, the slop is slightly reduced when Mesh B is used.

5.2 Effect of Heat Exchanges at the Wall

Simulations were then performed accounting for the conductive heat exchanges with the wall in the near-wall computing cell, for both the gas

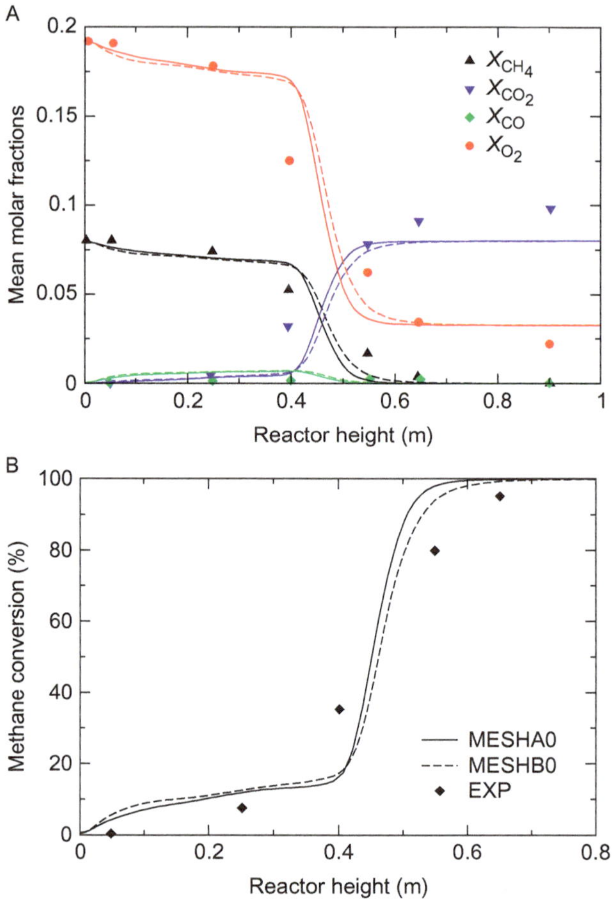

Fig. 12 Numerical, time-averaged, against experimental axial profiles. MESHA0: ——; MESHB0: - - - -. (A) Molar fractions. (B) Fuel conversion.

and the particulate phases. To do so, the following thermal boundary conditions were imposed:

$$\begin{cases} H_{k|_{\mathbf{x}=\mathbf{X}_0}} = H_k(T_W) & \text{if } z \leq 0.8\,\text{m}, \\ \nabla H_{k|_{\mathbf{x}=\mathbf{X}_0}} \cdot \mathbf{n} = 0 & \text{otherwise.} \end{cases}$$

The subscript $\mathbf{X}_0$ refers to the first grid point from the wall and T_W is the wall temperature, here imposed as $T_W = 707°\text{C}$.

The effect of the heat exchanges at the reactor wall can be analyzed by comparing the results obtained by MESHA1 and MESHB1 with that from MESHA0 and MESHB0, previously presented. The analysis of MESHA0 vs MESHA1 and that of MESHB0 vs MESHB1 led to similar conclusions,

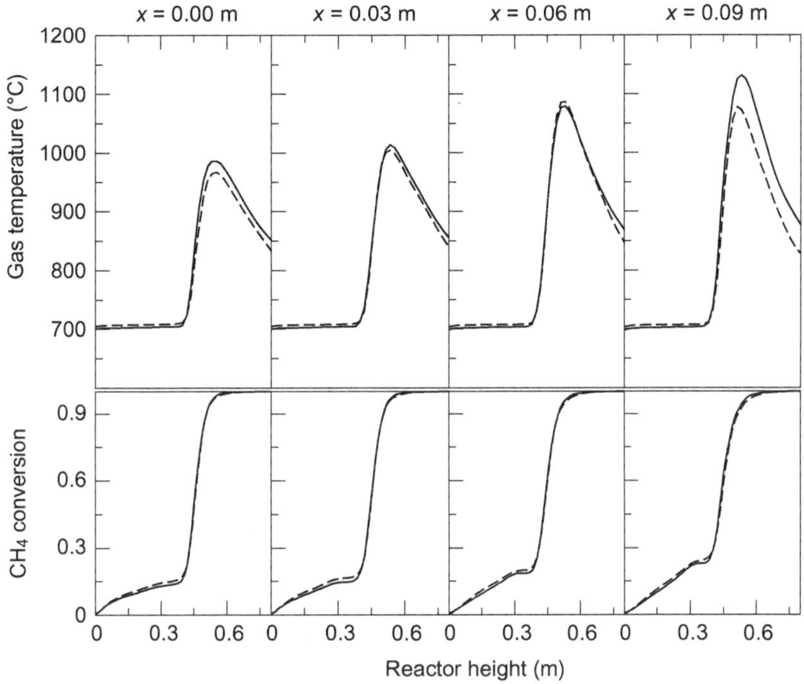

Fig. 13 Vertical profiles of the numerical time-averaged gas temperature and fuel conversion. MESHA0: ——; MESHA1: - - - -.

thus only the first one is here reported. Profiles of time-averaged gas temperature and methane conversion along the reactor height, at different x coordinates, are shown in Fig. 13. Globally, results show that taking into account the conductive heat exchange at the wall, by using Dirichlet conditions (for $z \leq 0.8$ m), only has a weak effect on the numerical predictions. However, conductive heat exchanges at the wall have to be taken into account in order to avoid a progressive increase of the gas and the particulate phase temperatures inside the bed, thus ensuring a thermodynamic statistically stationary regime for a long time simulation. In particular, previous studies showed that an important increase of the temperature into the bed occurs when finer grids are used without accounting for the conductive heat exchanges at the wall (Hamidouche et al., 2016). This is why such exchanges are further retained for finer meshes.

5.3 Effect of the Grid Refinement

Preliminary results gave the evidence that a better resolution of the bed hydrodynamics may produce a substantial effect on the predictions of the gas temperature and, by consequence, on the molar fractions as well. In this

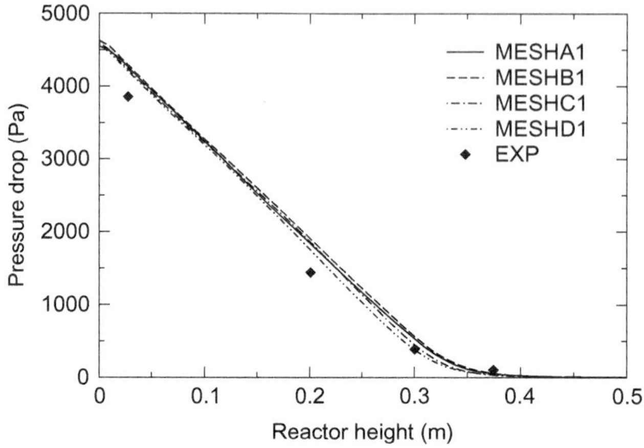

Fig. 14 MESHA1-D1: numerical, time-averaged, against experimental pressure drop at the wall ($x = 0.09$ m).

section, a mesh sensitivity analysis is carried out using Mesh A, Mesh B, and two more refined meshes, referred to as Mesh C and Mesh D (see Table 6). The simulations are then performed according to the conditions reported in Table 7. Fig. 14 shows the time-averaged pressure drop predictions for the simulations MESHA1, MESHB1, MESHC1, and MESHD1. Results show that the experimental results are globally well reproduced by all the simulations, while small differences are found comparing the four meshes at the top of the dense bed. Time-averaged gas temperature at the reactor center, as predicted by the four grids, are displayed in Fig. 15, compared with the experimental measurements. Improved results are obtained when finer grids are used. Results show that for the finest grid (Mesh D), the maximum of the gas temperature is drastically reduced reaching a value very close to that obtained in the experiments. Results also show that the location of the combustion zone is nearly the same for all the numerical simulations and, as previously mentioned, is slightly shifted toward the bed surface with respect to the experiments. Such a difference with the experimental measurements could be due to the chemical kinetics used in the numerical simulations, which is based on a simplified two-steps reaction mechanism instead of a complete reaction scheme. The two-steps mechanism is known to be very effective for lean mixtures in one-phase flows, within a range of temperatures close to that of the present study. However, its adequacy in dense fluidized beds should be further investigated. A parametric study carried out by Dufresne et al. (2016) on a one-dimensional static bed reactor, showed that, for bed temperatures lower than a critical value of about 800°C, the

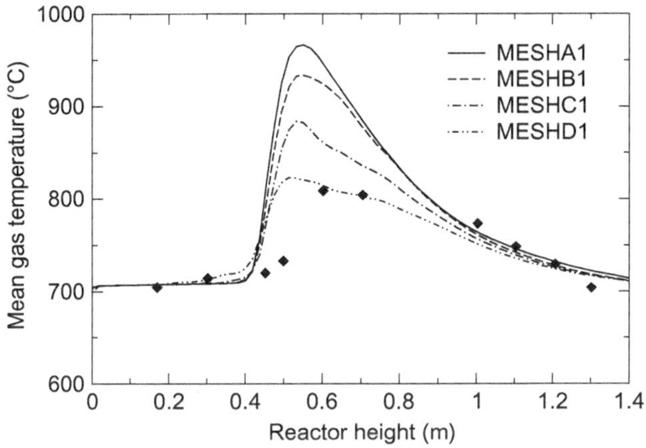

Fig. 15 MESHA1-D1: Numerical, time-averaged, against experimental gas temperature axial profiles.

Dryer and Glassman (1973) mechanism, in dilute regimes, leads to combustion zones slightly closer to the bed surface than that obtained with the detailed GRI (Gas Research Institute) mechanism for methane combustion (GRI-MECH 3.0, Smith et al., n.d.). On the other hand, previous studies (Sotudeh-Gharebagh and Chaouki, 2003 and references therein) have shown that the presence of inert solid particles may induce an increase of the homogeneous combustion ignition temperature. This effect may originate from free radicals' termination by the particle surfaces and should make the combustion zone to move far from the bed surface. In this case, it would be necessary to model the contribution of the heterogeneous processes in the combustion inhibition (Sotudeh-Gharebagh and Chaouki, 2003).

Globally, numerical results show that finer meshes make it possible to improve the predictions. It is thought that a mesh refinement allows the numerical simulation to better reproduce the hydrodynamics of the bed therefore improving thermodynamic predictions as well. Fig. 16 shows the time-averaged solid volume fraction and its variance as obtained at the reactor center from the different mesh refinements. Fig. 17 shows the time-averaged net solid mass flux and the upward one along the reactor axis as well. Results indicate mainly three different zones. A first zone, corresponding to $z < \sim 0.1$ m, in which particles are most concentrated. This zone corresponds to the zone in which bubbles of small size are formed. Bubbles then rise and grow into the bed before bursting at the bed surface. A second zone, corresponding to $\sim 0.1 \text{ m} < z < \sim 0.3$ m in which the particle concentration, at the reactor center, globally decreases. A third zone, for z

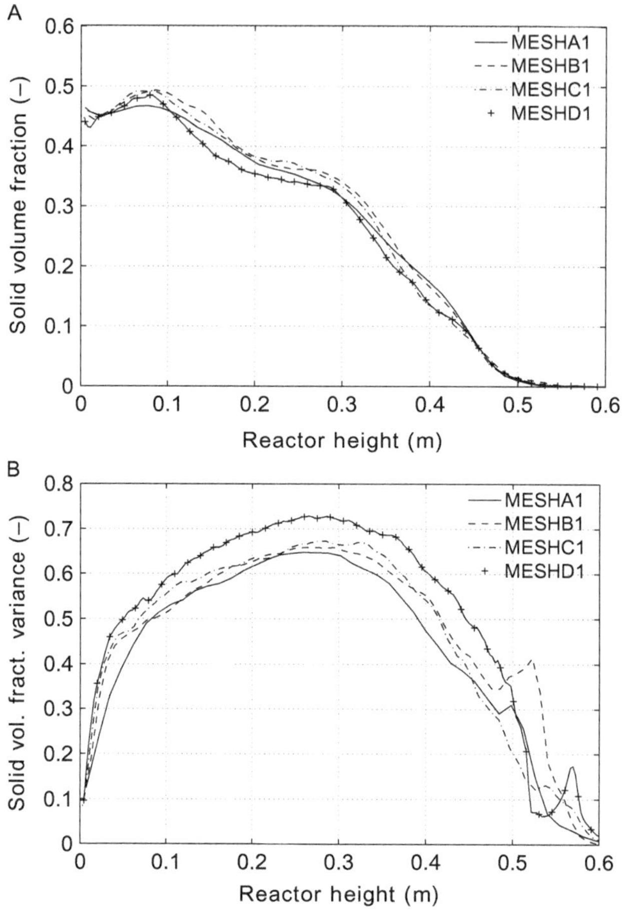

Fig. 16 Axial profiles of the time-averaged solid volume fraction $\overline{\alpha_p}$ (A) and solid volume fraction variance $\overline{\alpha_p'^2}/[\overline{\alpha_p}(\alpha_{p,max} - \overline{\alpha_p})]$ (B) for the numerical simulations MESHA1-B1-C1-D1.

$>\sim$0.3 m, in which such a decrease is even more pronounced. In the first zone (and slightly over for the coarser grids), the solid exhibits negative mass fluxes (Fig. 17A) meaning that the particles are caught in a recirculation zone which makes them going down, toward the bottom, at the reactor center. A positive and increasing net mass flux is instead observed corresponding to the second zone, in which particles go up, in the middle of the bed, toward the bed surface. Beyond it, the mass flux slowly decreases to zero in the center of the reactor. Particles indeed fall back to the bed close to the wall according to the well-known behavior, typical of the bubbling fluidized regimes. On the basis of the mass flux results (Fig. 17), a bed surface location corresponding to $z \sim 0.3$ m was identified for the finest mesh. This value

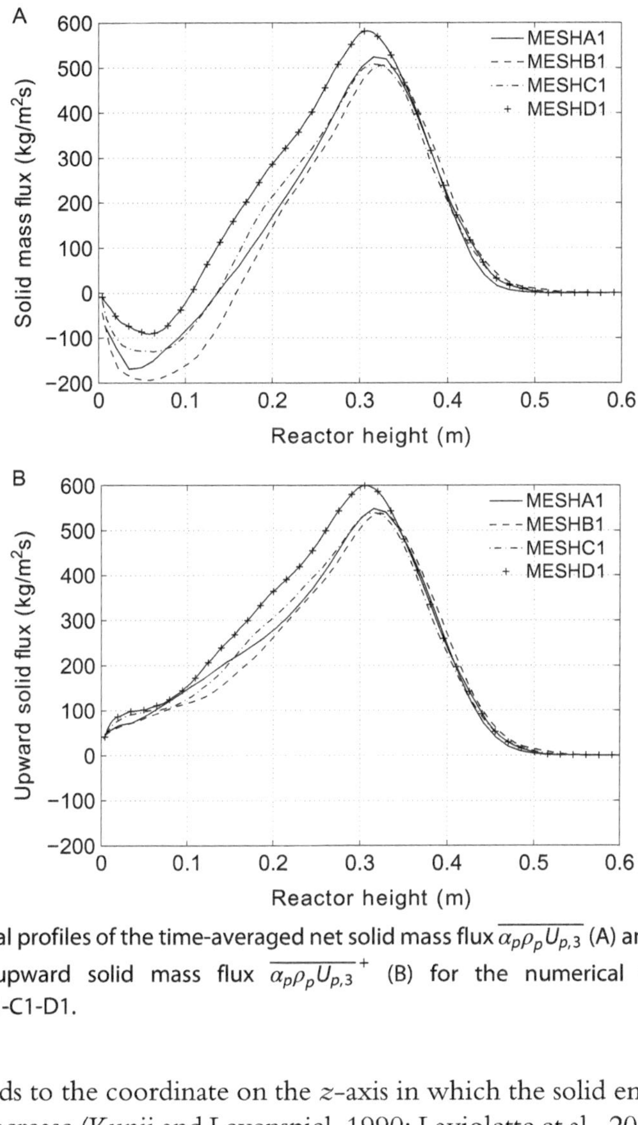

Fig. 17 Axial profiles of the time-averaged net solid mass flux $\overline{\alpha_p\rho_p U_{p,3}}$ (A) and the time-averaged upward solid mass flux $\overline{\alpha_p\rho_p U_{p,3}}^+$ (B) for the numerical simulations MESHA1-B1-C1-D1.

corresponds to the coordinate on the z-axis in which the solid entrainment starts to decrease (Kunii and Levenspiel, 1990; Laviolette et al., 2013). Based on the mass flux results, the reactor zone beyond $z \sim 0.3$ m, up to $z \sim 0.5$ m, is then referred to as freeboard in this work. Slightly higher bed surface locations are instead observed for the coarser meshes, as expected. Results displayed in Fig. 16B show the solid volume fraction variance normalized by the quantity $\overline{\alpha_p}(\alpha_{p,max} - \overline{\alpha_p})$ with $\alpha_{p,max} = 0.64$. This quantity represents the value of the variance into the bed in the limit case of a dense regime with $\alpha_p = 0$ into the bubbles and $\alpha_{p,max} = 0.64$ outside; the same may apply to the freeboard in the presence of strong clusters. Results show that for the finest grid, the solid volume fraction variance is higher into the bed and at the freeboard, indicating that particle spatial inhomogeneities are better reproduced when a finer mesh is used. Such inhomogeneities are the manifestation of a higher accuracy prediction of the bubbling zones into the bed and the bubble break-up at the freeboard. A finer mesh leads to better reproduce the gas pockets in the bed and their explosion at the bed surface.

The effect of the mesh size on the molar fraction and on the methane conversion is assessed by computing time-averaged quantities at the reactor center, as in the experiments. Results displayed in Fig. 18A and B, respectively, show that a refined grid improves the gas species predictions as well.

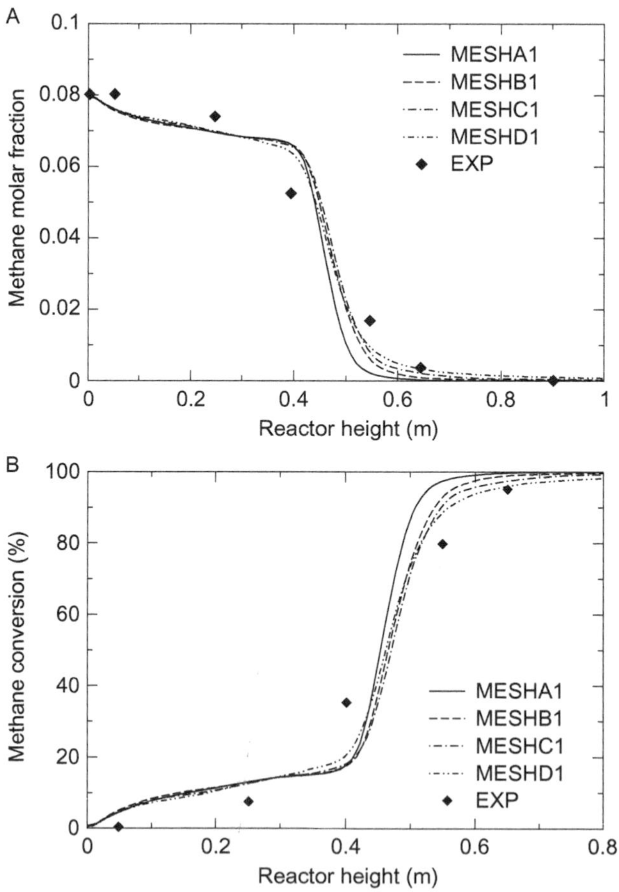

Fig. 18 MESHA1-D1: numerical, time-averaged, against experimental axial profiles. (A) Methane molar fraction. (B) Methane conversion.

5.4 Mesoscopic and Macroscopic Scale Analysis

In order to gain insight into the features of the natural gas combustion in fluidized beds, statistics are computed at macroscopic and mesoscopic level using, respectively, time-averaged and local-instantaneous fields. Correlations between two generic quantities A and B (which are either time-averaged or local-instantaneous fields) are computed using the correlation coefficient defined as follows:

$$C(A,B) = \frac{(\langle AB\rangle - \langle A\rangle\langle B\rangle)}{\sqrt{\langle A^2\rangle - \langle A\rangle^2}\sqrt{\langle B^2\rangle - \langle B\rangle^2}}, \tag{25}$$

where $\langle\cdot\rangle$ represents the spatial average computed in the whole reactor or in a selected zone within it. The average of A conditional on B is then:

$$\langle A|B = B'\rangle = \frac{\langle A\,\delta(B - B')\rangle}{\langle\delta(B - B')\rangle}. \tag{26}$$

This analysis aims to deepen and enhance the understanding of the gas–particle reactive flow in a dense regime. In particular, the effect of the bed hydrodynamics on the homogeneous (i.e., gas–gas) reactions should be better understood in order to provide information for the numerical modeling, allowing the CFD to give reliable predictions. In the previous section, it was shown that the finest mesh makes it possible to improve the predictions especially concerning the gas temperature. It was conjectured that such an improvement should depend on the ability of the finest mesh to predict the gas pockets into the bed and to reproduce their explosion at the bed surface. In the eruption zone, particles are indeed ejected at the (well-controlled) bed temperature and the heat exchanged with the hotter gas makes it possible to decrease the gas temperature. Because of such an interphase heat transfer, the temperature of the particles increases but this increase is very low compared to the gas temperature decrease. Under simplifying assumptions, this can be easily demonstrated on the basis of the energy conservation principle (involving the total (gas + particles) enthalpy) when in the presence of a high particle-to-fluid density ratio ($\rho_p/\rho_g \sim 8000$ at 700°C in the present study) and comparable gas and solid mass heat capacities.

Additionally, it was thought that a better prediction of the gas pockets into the bed would enhance the combustion process in the bed since the gas temperature inside the pocket is slightly higher than outside. Increasing the fuel conversion into the bed would lead to a lower amount of fuel at the freeboard and, by consequence, to a lower gas temperature in such a zone. In this case, in the bed, a higher amount of heat would be produced by the enhanced reaction, therefore increasing the gas temperature into the bed. This amount of heat should be removed through heat exchanges at the wall if one wish to maintain a well-controlled temperature and fuel conversion into the bed. When the heat is not removed, the gas temperature increases, in particular, in that zones of the bed in which the solid volume fraction is low. If the particle volume fraction is high enough, interphase heat transfers make it possible to reduce the gas temperature. At macroscopic scale, it is well known that the dispersed phase is more concentrated near the wall than at the reactor center (Ansart et al., 2017); a stationary near-wall higher solid distribution is indeed observed in fluidized beds (on the basis of time-averaged fields). This suggests that, in absence of heat exchanges at wall, the gas temperature would be larger at the center than at the wall. This is consistent with the results obtained from a previous study, revealing a substantial increase of the gas temperature into the bed center when a finer mesh is used and no conductive heat exchanges at the wall are take into account (Hamidouche et al., 2016). In order to investigate the effect of the gas pocket predictions on the dense reactive flow, a deeper analysis is carried out. Its results are presented and discussed below.

If the reaction is enhanced inside the gas pockets, then higher values of the methane conversion should be found in these zones. Fig. 19 shows the local and instantaneous fields of methane conversion, gas volume fraction, gas temperature, gas axial velocity, and methane mass change rate, as predicted by the finest mesh, at the end of the numerical simulation. Such fields are visualized in a slice in the middle of the reactor ($x = 0$). Results show that the gas temperature is indeed higher inside the gas pockets than outside, as well as the methane mass change rate, Ψ_{CH4}, in these zones. However, remarkably, results also show that the methane conversion is lower inside the pockets than outside. On the other hand, the gas exhibits higher ascending velocities inside the bubbles, suggesting that the conversion rate could not have enough time to be effective in such zones. An accurate estimate of the local residence time of the gas mixture in a selected zone of the reactor is not straightforward. An estimate of the mean gas residence time in the bed (of height ~0.3 m) may be obtained by using the interstitial mean gas velocity (Hesketh and Davidson, 1991). The latter is computed by dividing the fluidization velocity (here estimated as about 0.55 m/s at 700°C) by the void fraction (taken as ~0.55). According to this velocity, a corresponding mean residence time

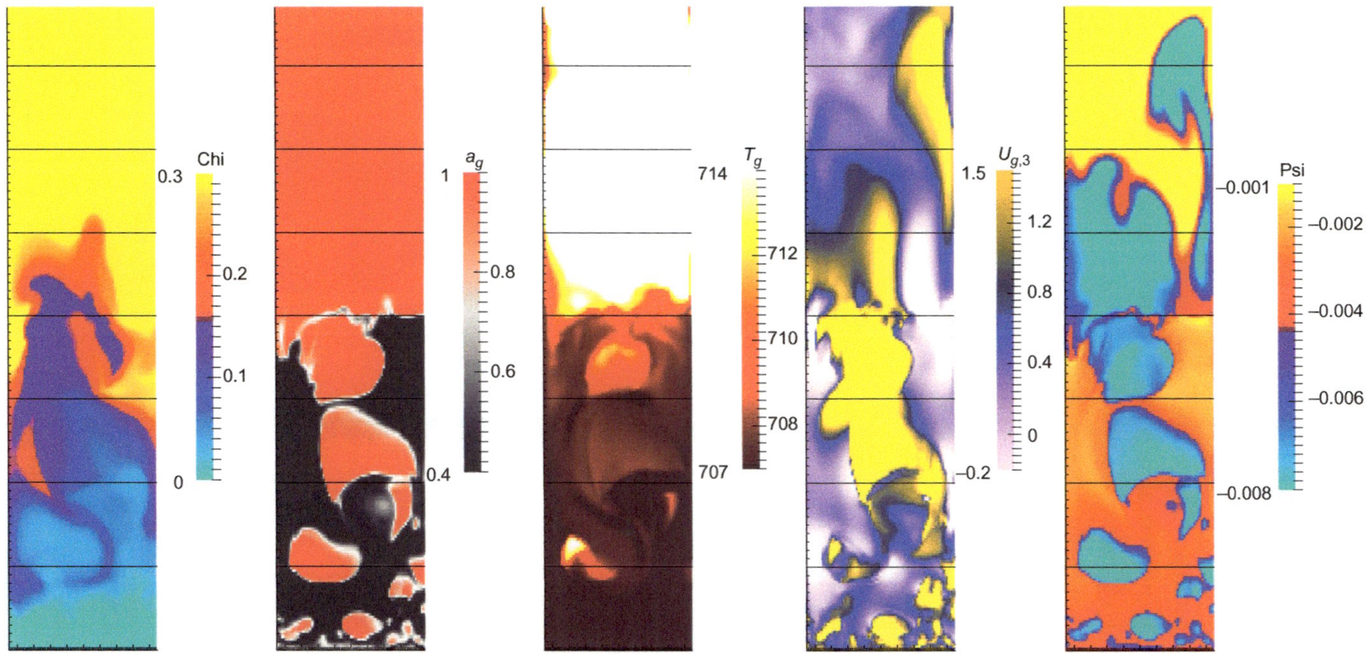

Fig. 19 Snapshots of instantaneous fields at $t = 80$ s for simulation MESHD1. From the *left* to the *right*: methane conversion, χ_{CH_4}, gas volume fraction, α_g, gas temperature, T_g (°C), gas vertical velocity, $U_{g,3}$ (m/s) and methane mass change rate, Ψ_{CH_4} (kg/(m^{3}s)). In the snapshots, the grid on z is shown in order to identify the axial zones into the bed. The distance between each grid line is 0.1 m.

of about 0.3 s is found. Otherwise, one can compute the mean gas velocity into the bed $\langle \overline{U_{g,3}} \rangle$ (also accounting for the distance from the center). This quantity spans from about 1.6 m/s in the center to 0.6 m/s close to the wall leading to mean residence times ranging between ~0.19 s and ~0.5 s. The reaction time can be instead estimated as follows: $\tau_{CH_4} = \rho_g \alpha_g Y_{CH_4}/|\Psi_{CH_4}|$. The local and instantaneous reaction time is computed for the finest mesh, at the end of the numerical simulation, and the result depicted in Fig. 20. Into the bed, this characteristic time scale is found to be quite homogeneous, corresponding to a mean value of about 2 s. Its value is found even higher when time-averaged quantities are used for its estimate (not shown). The ratio between the bed height and the vertical gas velocity is also shown in Fig. 20, for a comparison purpose. Such a ratio has dimensions of a time even

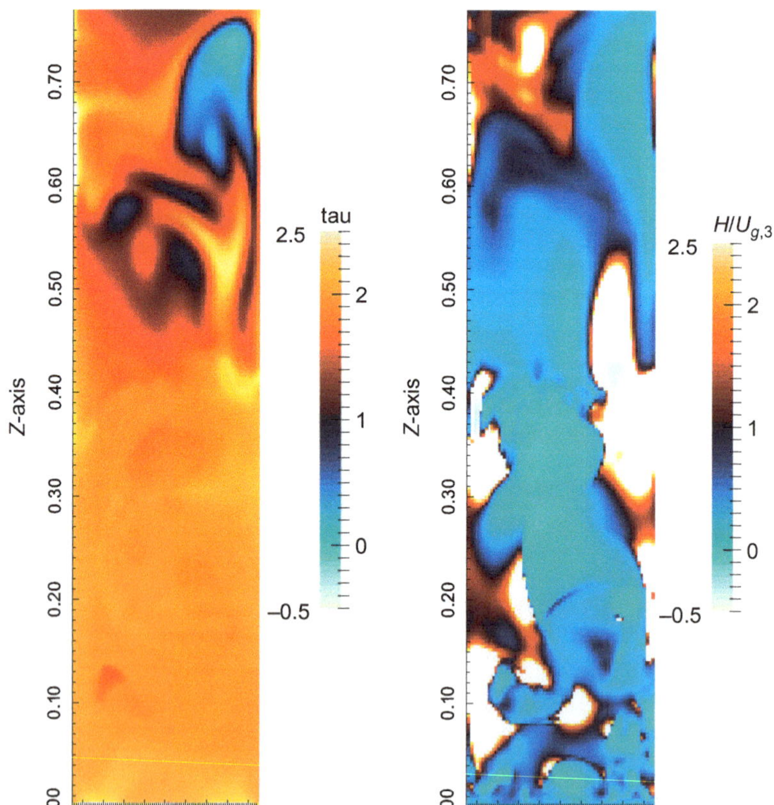

Fig. 20 Snapshots of the instantaneous (t = 80 s) reaction time (*left*) and of the ratio between the bed height (~0.3 m) and the instantaneous vertical gas velocity (*right*), for simulation MESHD1.

if it is not. Negative values indeed appear due to the negative gas velocities. They were intentionally left in order to highlight the zone where the gas is moving down and the residence time could be enhanced. On a speculative basis, this quantity is also compared with the local and instantaneous estimate of the reaction time. Globally, results suggest that the competition between characteristic time scales may be a key point in understanding the fuel conversion in the reactor. This question is not new and already discussed in the literature (see, for instance, Hesketh and Davidson, 1991). Results show, indeed, that the reaction time may be higher than the residence time, especially, in those zones where the gas velocity is high (typically inside the bubbles).

Based on these results, it is believed that the fuel conversion does not occur neither into the pockets nor close to their boundary surface, in the bed, but rather in those zones where the residence time is propitious to allow the chemical reaction to take place. Instantaneous results localize these zones at the freeboard and near the wall, into the bed, close to the bed surface. Considering the selected operating point, combustion should occur in an intermediate regime, according to the definition given by Żukowski (2003).

In order to understand if such conclusions may be generalized at macroscopic scale, time-averaged quantities are investigated. Fig. 21 shows the same fields than Fig. 19, averaged in time. Results confirm a larger fuel conversion in the upper part of the bed, close to the wall (~ 0.2 m $\leq z \leq \sim 0.3$ m), and at the freeboard ($z > \sim 0.3$ m). Results also show that, in the first part of the freeboard (~ 0.3 m $\leq z \leq \sim 0.4$ m), the methane mass change rate is larger at the center than close to the wall as well as the gas temperature in such a zone. The latter corresponds to the eruption zone where particles are ejected from the bed center and fall back to the bed, close to the wall. Lower gas concentrations are indeed found nearby the reactor walls, in the upper part of the bed and in the lower part of the freeboard. In these zones, negative vertical gas velocities are also observed. Globally, results show that the reactor sector close to the bed surface is a crucial zone in the reactive dense fluidized bed; this zone should be accurately reproduced in order to provide reliable predictions. Such a zone is further investigated comparing two different grid refinements. Fig. 22 shows the time-averaged methane conversion for the simulations MESHB1 and MESHD1 at the station along the axial direction corresponding to $z =$ 0.3 m. Results show that the methane conversion is higher close to the wall, for the coarser mesh as well. Furthermore, close to the wall, the methane conversion is even higher when the finest grid is used. Higher methane

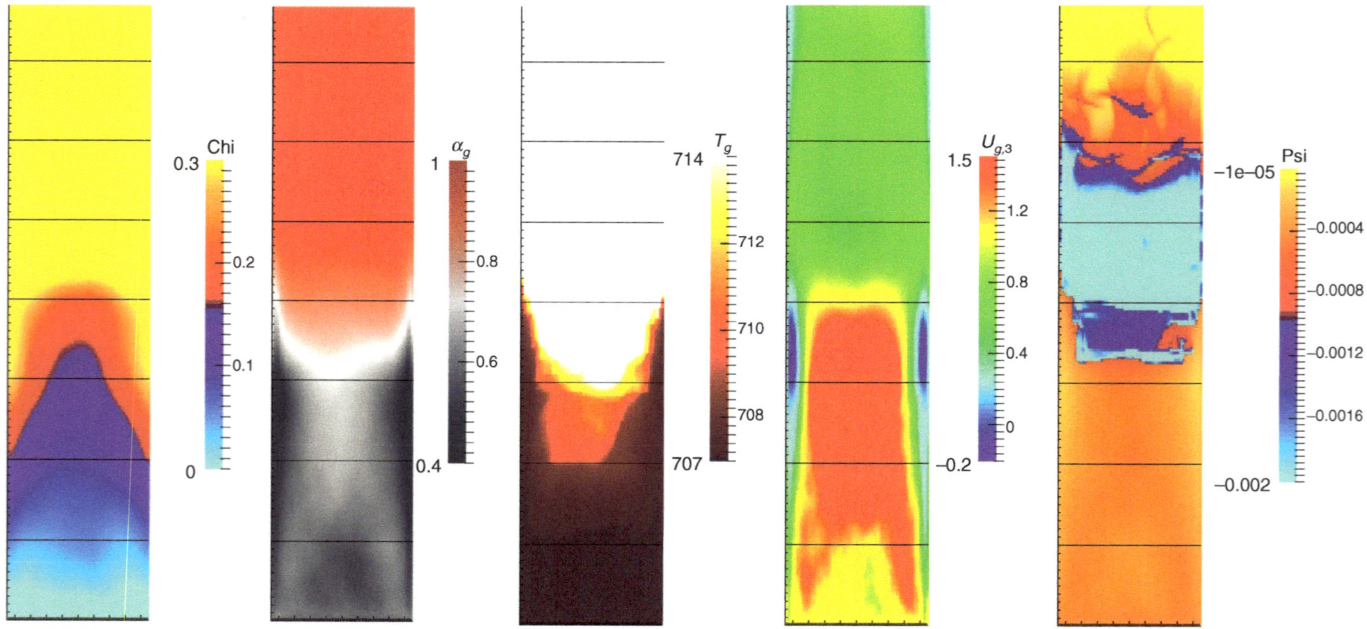

Fig. 21 Time-averaged fields for simulation MESHD1. From the *left* to the *right*: methane conversion, χ_{CH_4}, gas volume fraction, α_g, gas temperature, T_g (°C), gas vertical velocity, $U_{g,3}$ (m/s), and methane mass change rate, Ψ_{CH_4} (kg/(m^{3}s)).

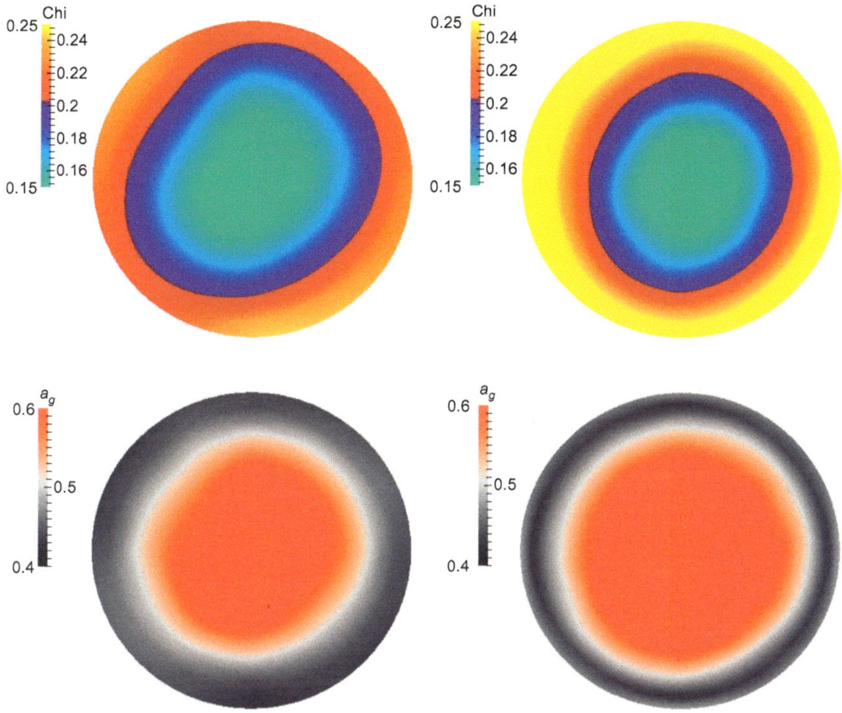

Fig. 22 Snapshots of the time-averaged methane conversion (*top*) and gas volume fraction (*bottom*) over the plane $z = 0.3$ m. MESHB1 (*left*) and MESHD1 (*right*).

conversion does not correspond, however, to higher gas temperature at the same location, as it can be seen by Fig. 23. The latter shows indeed that for MESHD1 the highest gas temperature is at the center of the reactor where the lowest methane conversion is found. The analysis at mesoscopic scale is thus confirmed at macroscopic scale. It is also worth of note that, at this location, the gas mixture exhibits important negative time-averaged vertical velocities near the wall for MESHD1 unlike MESHB1.

In order to understand what could be at the origin of a greater conversion near the wall, the bed hydrodynamics is further investigated. Fig. 24 shows the time-averaged particle velocity field as obtained in a slice in the middle of the reactor ($x = 0$). Arrows are then colored by the methane conversion. Results show that particles essentially move upward in the reactor center and downward near its walls, as well known. Results also show that in the lower part of the freeboard, close to ejection zone, particles are caught in a toroidal loop which promotes mixing from the top and from the center toward the wall. The loop center is localized inside the eruption zone at

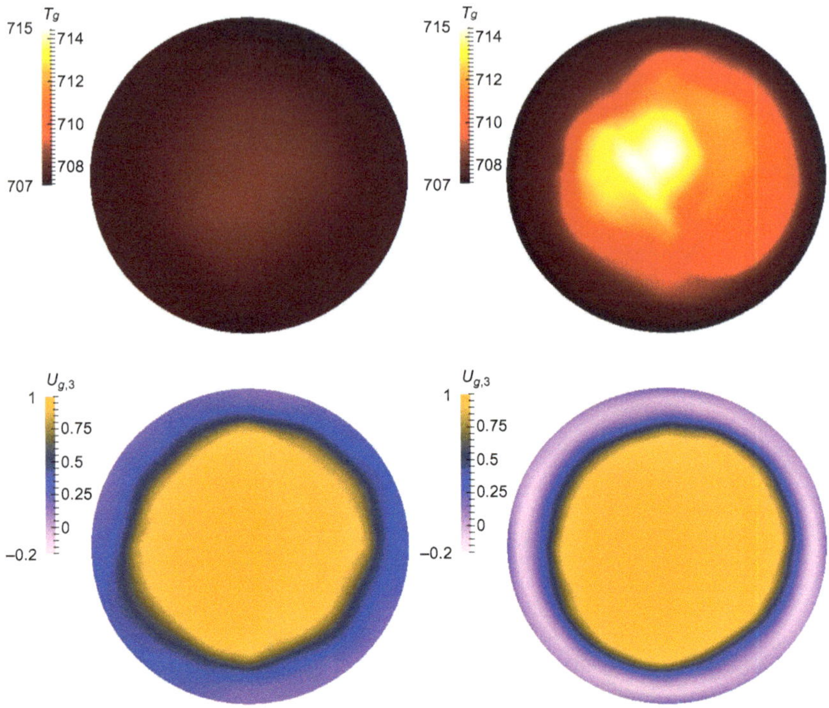

Fig. 23 Snapshots of the time-averaged gas temperature (°C) (*top*) and gas vertical velocity (m/s) (*bottom*) over the plane $z = 0.3$ m. MESHB1 (*left*) and MESHD1 (*right*).

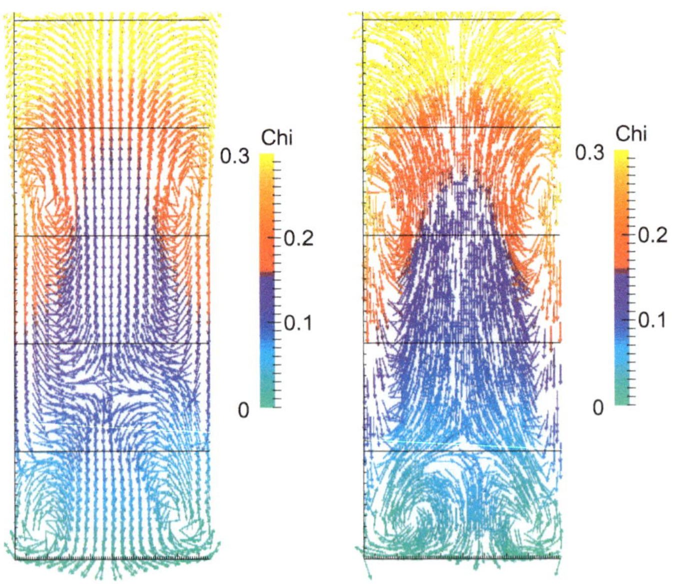

Fig. 24 Time-averaged particle velocity field on a slice in the middle of the reactor ($x = 0$ m). MESHB1 (*left*) and MESHD1 (*right*). The arrow scale factor (0.02) is the same for both the meshes as well as the number of random points (5000).

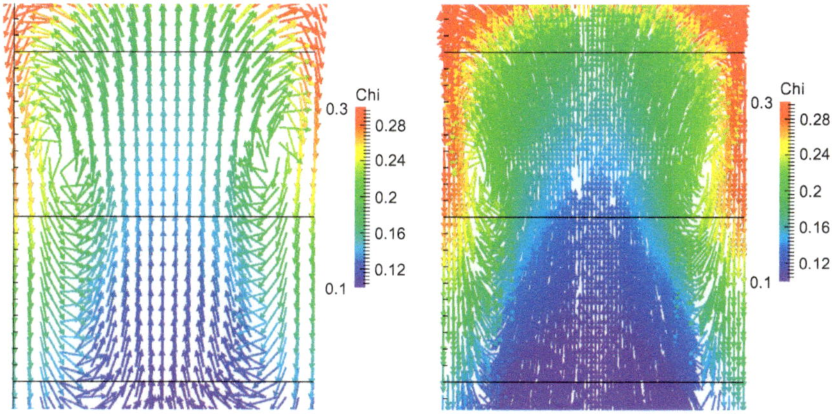

Fig. 25 Zoom of the time-averaged particle velocity field on a slice in the middle of the reactor ($x = 0$ m), including zones in which $z \in [0.2 - 0.4]$ m. MESHB1 (*left*) and MESHD1 (*right*). The arrow scale factor (0.02) is the same for both the meshes, the number of random points is increased to 15,000 for MESHD1.

$z \sim 0.33$ m. Such a loop, as well as a counter-wise loop at the bottom of the reactor, were already observed by Fede et al. (2016) who investigated the solid wall boundary conditions in 3D numerical simulations of a lab-scale pressurized dense fluidized bed. The authors pointed out the effectiveness of the no-slip wall boundary conditions, the same used in the present work, in accounting for the wall roughness effect on the particle hydrodynamics.

Concerning the effects of the hydrodynamics on the reaction, noteworthy is that, inside the bed, a larger fuel conversion is found corresponding to the solid recirculation zones (a more detailed view is given in Fig. 25). Results suggest that the hydrodynamic behavior of the solid has a strong effect on the gas combustion, as expected. Results also indicate that hydrodynamics and reactions are strongly inhomogeneous inside the bed, at the macroscopic scale as well, and that statistics should be computed over selected zones rather than over the entire bed, in order to capture all the bed peculiarities.

For this purpose, hereafter, quantities are computed and analyzed over selected zones. Fig. 26 shows the radial profiles of the mean vertical gas and particle velocities and of the mean methane molar fraction, in a layer including the toroidal loop center, for the simulations MESHB1 and MESHD1. Inside the layer, mean are computed by spatially averaging the time-averaged quantities over rings of width comparable to the relative mesh cell size. Each mean is then displayed versus the center of the corresponding bin. For both the simulations, results show that the mean

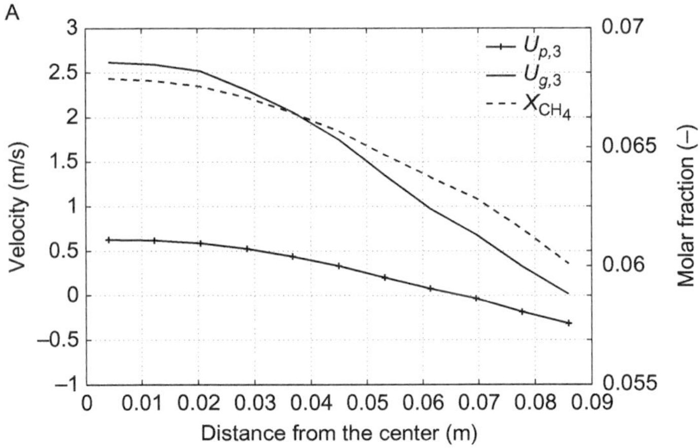

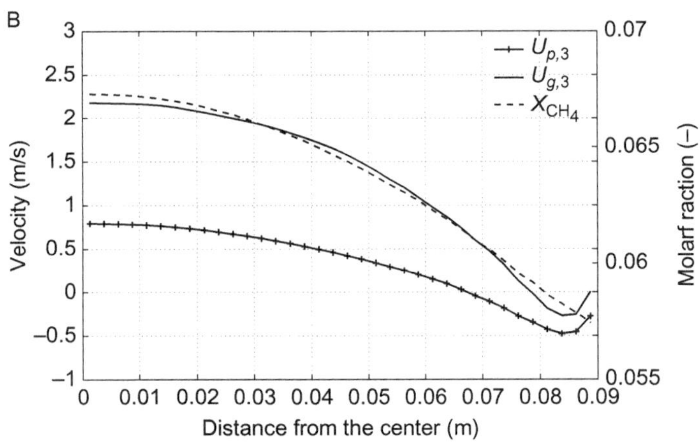

Fig. 26 Radial profiles of the mean (time and spatial averaged) vertical gas and particle velocities and methane molar fraction in the layer $z \in [0.32, 0.34]$ m. MESHB1 (A): 11 bins of equal size 8.18×10^{-3} m and MESHD1 (B): 36 bins of equal size 2.5×10^{-3} m.

methane molar fraction is larger at the center of the reactor than at the reactor wall, consistently with what was previously observed for the time-averaged methane conversion. Results also show that the mean vertical particle velocity is negative close to the wall, confirming the downward movement of the particles in such a zone.

Further, looking at the distance from the wall, results suggest a similar loop extension for both the simulations, also confirmed at several other locations (not shown). However, the finest mesh exhibits a nonmonotonic trend of the particle velocity near the wall and, moreover, negative vertical gas velocities within the toroidal loop (as already observed). In order to quantify the occurrence of negative gas velocities, the spatial occurrence density

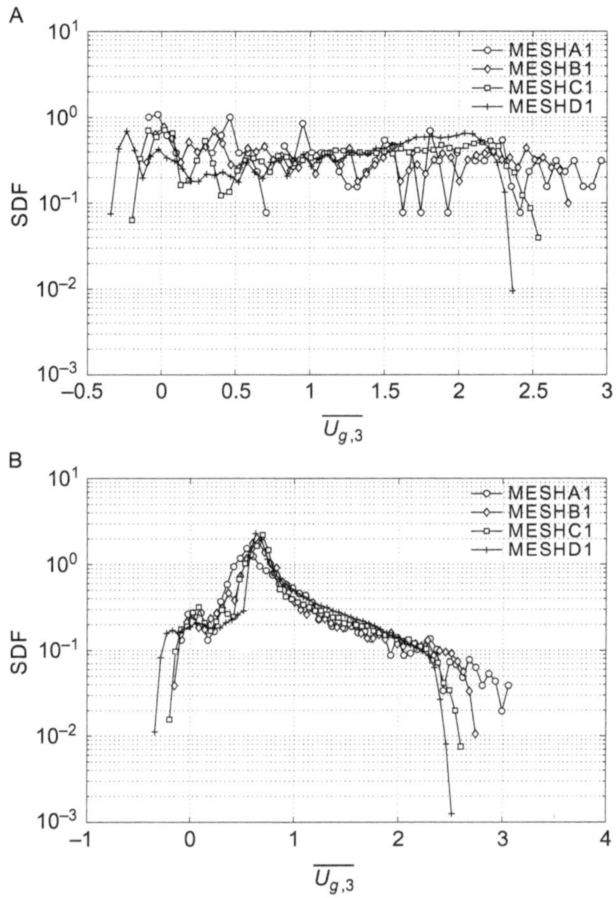

Fig. 27 Spatial occurrence density function of the time-averaged axial gas velocity (m/s), for $z \in [0.32, 0.34]$ m (A) and within an extended zone for $z \in [0.30, 0.50]$ m (B), for the numerical simulations MESHA1-B1-C1-D1.

function (SDF) of the time-averaged vertical gas velocity is computed, within the same selected layer and within an extended layer, $z \in [0.30, 0.50]$ m; results are displayed in Fig. 27. Negative gas values are found for all the meshes; however, MESHD1 exhibits the larger extent.

Results showed by Fig. 26, also point out that the mean methane molar fraction predicted by MESHD1 is smaller than the one obtained from MESHB1, especially in that zone corresponding to the loop extension at this axial location. The loop extension (here referred to the transversal recirculation length) is found to be $\approx \pm 0.02$ m for both the simulations. Within the same selected layer, mean solid volume fraction and mean solid fluxes are

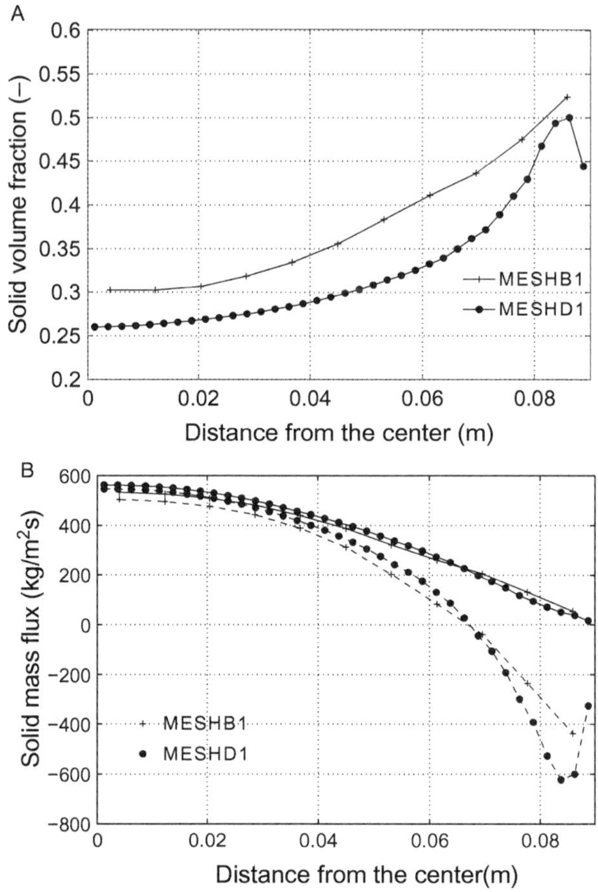

Fig. 28 Radial distribution of the mean (time and spatial averaged) solid volume fraction (A) and solid mass fluxes (B) in the layer $z \in [0.32, 0.34]$ m. In (B) *solid line* corresponds to $\overline{\alpha_p \rho_p W_p}^+$ and *dashed line* to $\overline{\alpha_p \rho_p W_p}$.

computed and results depicted in Fig. 28. The radial distributions confirm the aforementioned observations and give the evidence of a different resolution of MESHB1 compared to MESHD1, especially near the wall.

At this point, it is believed that the bed hydrodynamics affects the reaction depending on the predictions of both the particulate and the gas-phase behaviors. Concerning the latter, it is thought that the aforementioned negative gas velocities observed near the wall play an important role on the fuel conversion. Fig. 29 shows streamlines generated using the time-averaged gas velocity field on a slice in the middle of the reactor ($x = 0$ m). For both the numerical simulations, results show that within the toroidal loop a stagnation

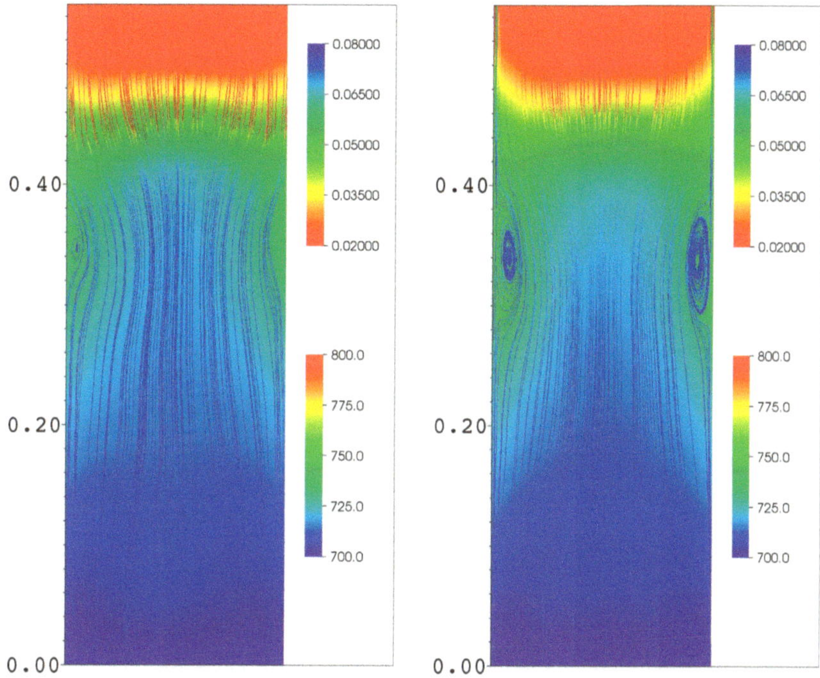

Fig. 29 Streamlines generated from the time-averaged gas velocity field on a slice in the middle of the reactor ($x = 0$ m) using 100 random points. MESHB1 (*left*) and MESHD1 (*right*). Color bars: methane molar fraction (*top*), gas temperature (°C) (*bottom*).

zone exists where the gas may reside for a long time. Results also indicate that for MESHD1, inside the loop, the gas is convected in a recirculation zone which should lead to residence times even longer than for MESHB1. In Fig. 29, the streamlines are colored by the mean gas temperature while the background contour gives the mean methane molar fraction. Results suggest that gas temperatures and fuel molar fractions are not (or few) correlated in the zone corresponding to the recirculation loop, while they are outside.

In order to investigate this point, conditional averages and correlation coefficients are computed. The reactor is divided in four zones denoted as Zone 1 ($z \leq 0.3$ m), Zone 2 (0.3 m $< z \leq 0.4$ m), Zone 3 (0.4 m $< z \leq 0.5$ m), and Zone 4 (0.50 m$< z \leq 1.4$ m); the disengagement zone is discarded in this analysis. The freeboard, identified as the reactor locations corresponding to 0.3 m $< z \leq 0.5$ m, is therefore split in two regions in order to capture the different peculiarities of the flows over the bed. In each zone, the conditional average $\langle \overline{\chi_{\mathrm{CH_4}}} \mid \overline{T_g} \rangle$ and the correlation coefficient $C(\overline{\chi_{\mathrm{CH_4}}}, \overline{T_g})$ are computed and results depicted in Figs. 30 and 31,

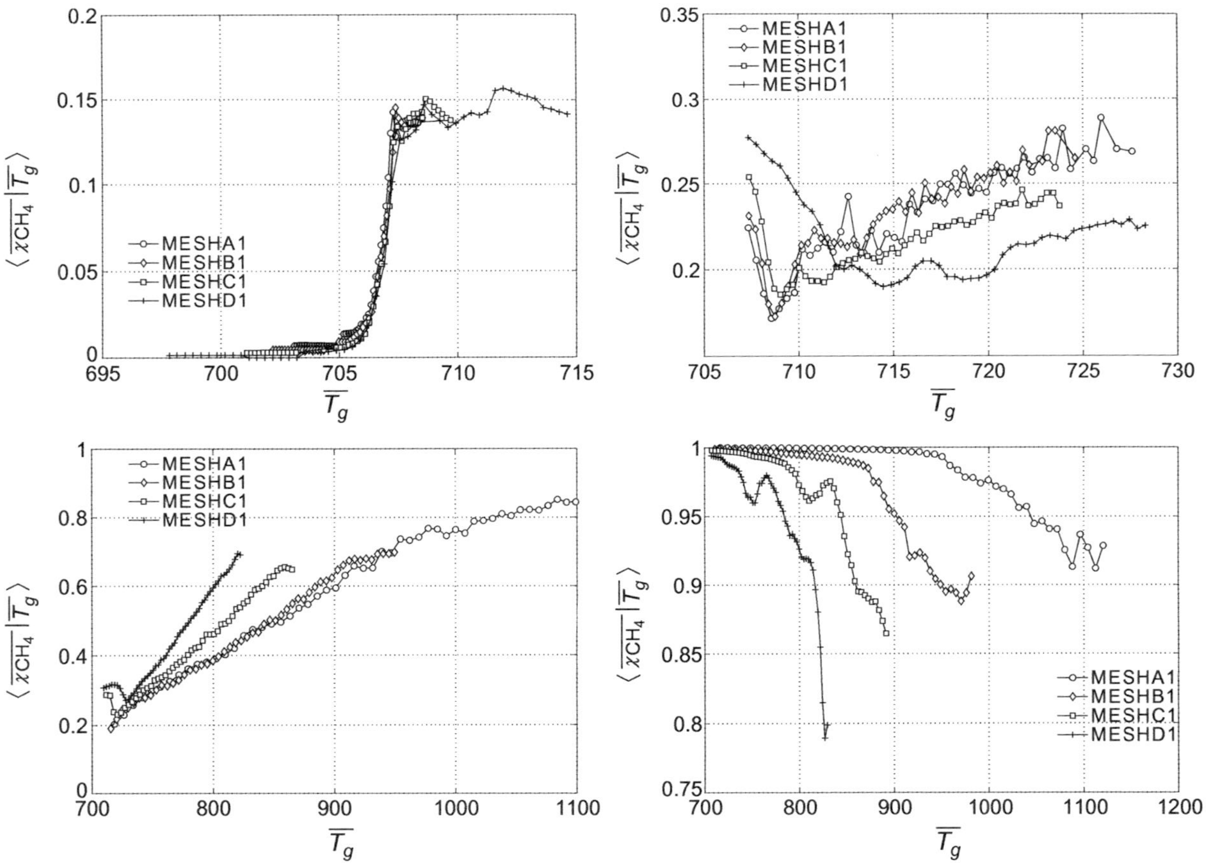

Fig. 30 Spatial average of the time-averaged methane conversion conditional on the time-averaged gas temperature (°C) in the four zones of the reactor for simulations MESHA1-B1-C1-D1. *First row*: Zone 1 (*left*) and Zone 2 (*right*). *Second row*: Zone 3 (*left*) and Zone 4 (*right*).

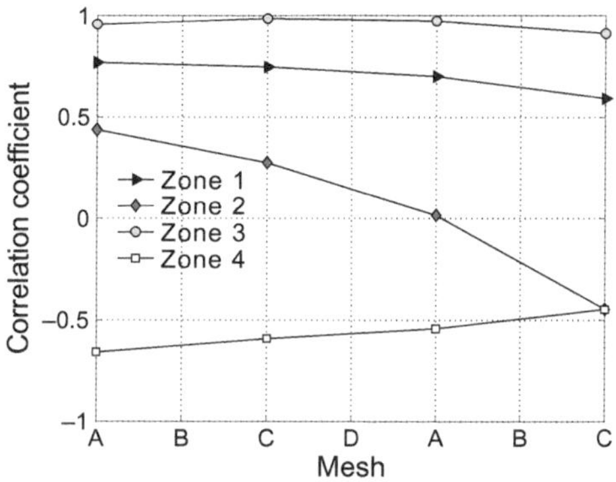

Fig. 31 Correlation coefficients of the time-averaged methane conversion with the time-averaged gas temperature in the four zones of the reactor for simulations MESHA1-B1-C1-D1.

respectively. Results show that the mean methane conversion is quite well correlated with the mean gas temperature in Zone 1 and strongly correlated with it in Zone 3. As expected, smaller correlations are instead observed in Zone 2 and, most important, they worsen with the grid refinement. In Zone 3, results show a greater extent of the fuel conversion when coarser grids are used, leading to higher temperature predictions. In this zone, a linear increasing of the mean fuel conversion with the temperature is observed. This is attributed to the range in which temperature is spanning. This range represents indeed the first part of the exponential Arrhenius law where the reaction rate slowing increases. Negative correlations are instead observed in Zone 4. In this zone, methane is completely burning but the gas temperature decreases because of the radiative heat transfers at the wall. Conditional averages of the time-averaged methane conversion on the time-averaged gas volume fraction are also computed and results depicted in Fig. 32. Results show that, at macroscopic scale, the methane conversion increases with the fluid volume fraction in the upper part of the freeboard and over (Zone 3 and 4). Results indicate instead a more complex behavior inside the bed and above the bed surface (Zone 1 and 2). In particular, two different trends are observed: the fuel conversion decreases with the gas volume fraction for lower values of α_g and increases with the gas volume fraction for higher values of it. In order to statistically quantify this behavior, a joint SDF, $f(\overline{\alpha_g}, \overline{\chi_{CH_4}})$, is computed within the Zone 2 and results depicted in Fig. 33A. Results show that most of the occurrences correspond to high gas volume fractions and moderate fuel conversions. Results also indicate a greater fuel conversion in highly concentrated particle zones. The latter should correspond to the

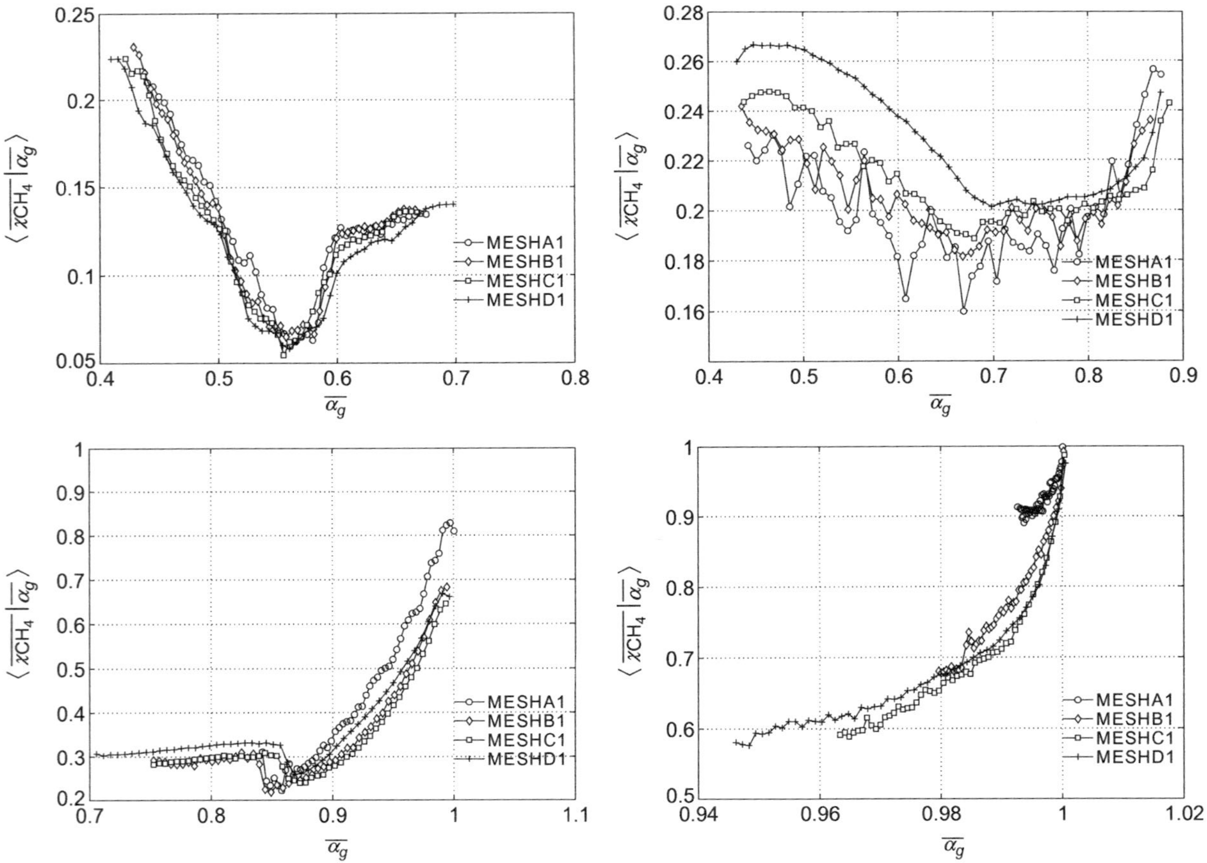

Fig. 32 Spatial average of the time-averaged methane conversion conditional on the time-averaged gas volume fraction in four zones of the reactor for simulations MESHA1-B1-C1-D1. *First row*: Zone 1 (*left*) and Zone 2 (*right*). *Second row*: Zone 3 (*left*) and Zone 4 (*right*).

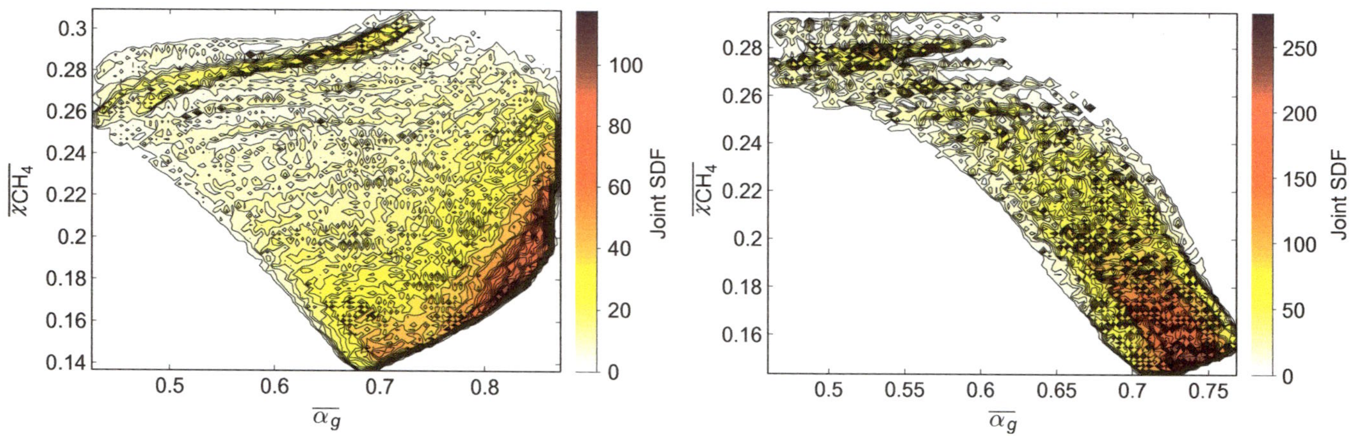

Fig. 33 Joint SDF of the time-averaged methane conversion and the time-averaged gas volume fraction in Zone 2 (A) and in the zone corresponding to $z \in [0.32, 0.34]$ (B), for simulation MESHD1.

zones close to the wall where the gas flows with the particles falling back to the bed. In such zones, a leaner mixture recirculate inside the toroidal loop. The fuel conversion is higher because the gas comes from the freeboard and backs to the bed, and it may even increase because of an increased residence time which promotes combustion. In these zones, the gas temperature is controlled by the interphase heat exchange with the highly concentrated particulate phase. Fig. 33B shows the joint SDF in the zone restricted to the loop center ($z \in [0.32, 0.34]$); results confirm the aforementioned observations.

5.5 Evaluation of the Subgrid Turbulent Mixing Effect on the Combustion Process

The analysis presented in the previous sections has shown that a good resolution of the gas–solid mixture hydrodynamic in and above the bed leads to improve thermodynamic (temperature) and chemistry (molar fraction) predictions for the gaseous mixture. This resolution depends on the mesh refinement. Such a refinement not only has an impact on the thermodynamics through indirectly the hydrodynamic resolution but also through directly the local mixing predictions. With the aim to model the effect of the subgrid turbulent mixing on the combustion, a finite rate chemistry/eddy dissipation approach was used.

In most cases, gas-phase chemical reactions are very fast and the rate of combustion is more influenced by the turbulent mixing than by chemical kinetic features. In those cases, the combustion rate can be determined by the mixing rate between the reactant and the product species, transported by the turbulent eddies. Based on such an idea, Spalding (1971, 1977) proposed, for a premixed combustion, to express the averaged reaction rate as a function of a turbulent mixing time, modeled as the ratio between the fluid turbulent kinetic energy and its dissipation rate, and the variance of the fuel mass fraction. This model, known as Eddy Break-Up (EBU) model, cannot make any difference between rich and lean mixtures. A model, known as Eddy Dissipation Model (EDM), was proposed by Magnussen and Hjertager (1977), who replaced the variance of the fuel mass fraction by the mean mass fraction of the deficient species (fuel for lean or oxygen for rich mixtures). Further, Magnussen (1981, 2005) extended this model and proposed the Eddy Dissipation Concept (EDC). The author argued that the reaction space is provided by the so-called fine structures, where the

Table 10 Combustion Modeling Used in the Numerical Simulation MESHA2

Dryer and Glassman Mechanism (DG)		**Eddy Dissipation Approach (ED)**	
Chemical reaction time		Turbulence mixing time	
$\tau_{ch,1}=\frac{[CH_4]}{R_{ch,1}}$	$\tau_{ch,2}=\frac{[CO]}{R_{ch,2}}$	$\tau_{mix}=A\left(\frac{1}{2}S^*_{g,ij}S^*_{g,ij}\right)^{-1/2}$ with $A=1$	
Chemical reaction rate		Turbulence mixing rate	
$R_{ch,1}$	$R_{ch,2}$	$R_{mix,1}=\frac{[CH_4]}{\tau_{mix}}$	$R_{mix,2}=\frac{[CO]}{\tau_{mix}}$
Eq. (10)	Eq. (11)		
Resulting DG/ED Model			
	Combustion process		
(R_1)	$CH_4+3/2\ O_2\rightarrow CO+2H_2O$		
(R_2)	$CO+1/2\ O_2\rightarrow CO_2$		
Resulting reaction times		Resulting reaction rates	
$\tau_1=\max(\tau_{ch,1};\ \tau_{mix})$		$R_1=\min(R_{ch,1};\ R_{mix,1})$	
$\tau_2=\max(\tau_{ch,2};\ \tau_{mix})$		$R_2=\min(R_{ch,2};\ R_{mix,2})$	

energy dissipation and the molecular mixing occur. In the present work, in order to evaluate the effect of the subgrid turbulent mixing on the combustion, an eddy dissipation approach, based on the deviatoric rate of strain ($S^*_{g,ij}$), is used (Xue and Fox, 2014). According to Xue and Fox (2014), the gas reaction rate is modeled as the minimum between the chemical kinetic rate R_{ch} and the turbulent mixing rate R_{mix}, namely $R_{min}=\min(R_{ch}, R_{mix})$. The chemical kinetic rate, R_{ch}, is defined using the finite rate chemistry model proposed by Dryer and Glassman (1973), given in Section 3.2. A summary of the quantities involved in the modeling is given in Table 10.

The model was used in conjunction with the coarsest grid (Mesh A) and the numerical simulation referred to as MESHA2 (see Table 7). The latter is then compared with MESHA1 based on the Dryer and Glassman (1973) kinetics (DG). For both the simulations MESHA1 and MESHA2, axial profiles of the time-averaged gas temperature and mean molar fractions at the reactor center are shown in Figs. 34 and 35, respectively. Results show that inside the bed, MESHA2, which uses the DG/ED model for the combustion, predicts very similar quantities compared with the simulation MESHA1, while in the freeboard region and above results are substantially different. In particular, it appears that MESHA2 overestimates the gas temperature even more than MESHA1. In fact, the predictions are not

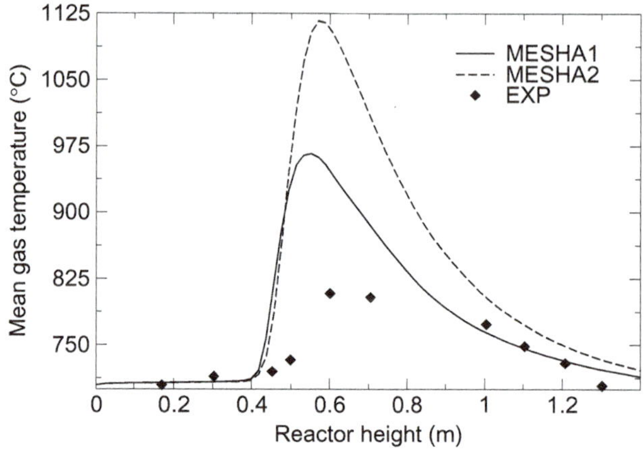

Fig. 34 Time-averaged gas temperature profiles. MESHA1 and MESHA2 vs experiments.

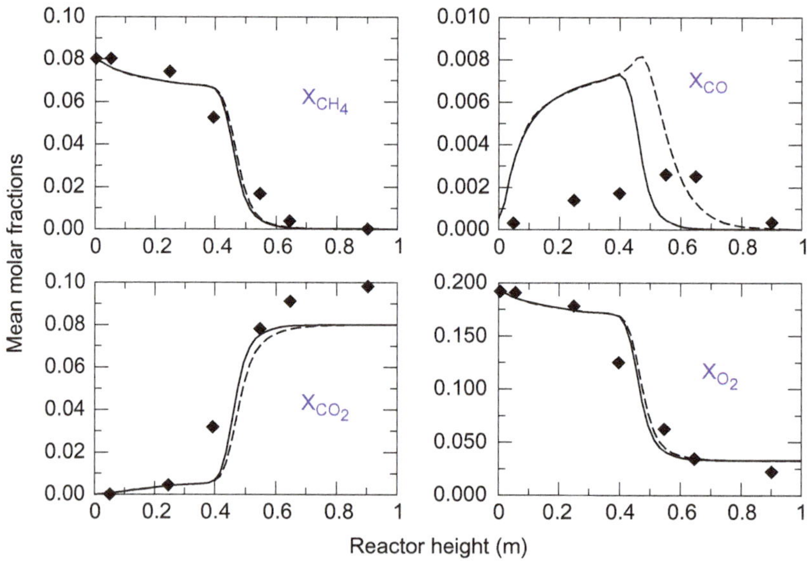

Fig. 35 Time-averaged molar fraction profiles. MESHA1 (——) and MESHA2 (- - - -) vs experiments (◆).

improved by the DG/ED model as instead expected. Such results suggest that, inside the bed, the overall combustion process is controlled by the chemical reaction while, in the freeboard region, chemical reaction, and subgrid turbulent mixing are competitive mechanisms.

In order to investigate this point, the "probability" to find the chemical reaction ($\mathcal{P}_{ch}$) or the subgrid turbulent mixing ($\mathcal{P}_{mix}$) as limiting mechanism, for each reaction step (R_1) and (R_2), is computed as follows:

$$\mathcal{P}_{ch} = \mathcal{P}(R_{ch} \leq R_{mix}) \text{ and } \mathcal{P}_{mix} = \mathcal{P}(R_{ch} > R_{mix}) \text{ with } \mathcal{P}_{ch} + \mathcal{P}_{mix} = 1.$$

A unity value of $\mathcal{P}_{ch}$ means that the overall process is dominated by the chemical reaction, whereas if $\mathcal{P}_{mix} = 1$, the overall process is then dominated by the subgrid turbulent mixing. Results obtained from simulation MESHA2 are displayed in Fig. 36A and B for the reactions R_1 ($CH_4 \rightarrow CO$) and R_2 ($CO \rightarrow CO_2$), respectively. As expected, inside the bed, the subgrid turbulent mixing does not affect the combustion process. Therefore, in this region, the gas temperature and the species molar fractions are the same for MESHA2 and MESHA1. Noteworthy, same results are also observed in the freeboard region close to the bed surface (Zone 2). Beyond $z \sim 0.4$ m, the probabilities

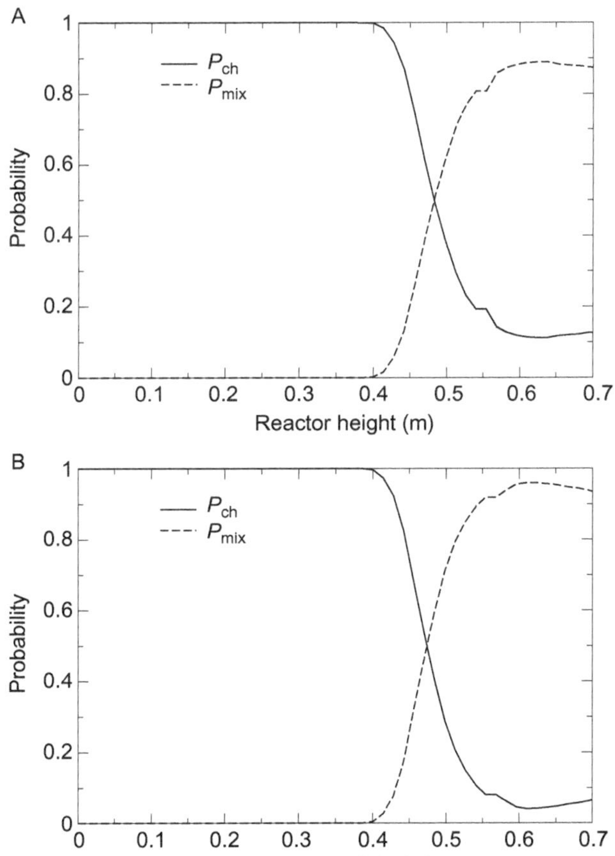

Fig. 36 MESHA2: Chemical reaction versus turbulence mixing on the overall combustion process: (A) reaction R_1 ($CH_4 \rightarrow CO$) and (B) reaction R_2 ($CO \rightarrow CO_2$).

$\mathcal{P}_{ch}$ and $\mathcal{P}_{mix}$ become comparable, confirming that chemical reaction and subgrid turbulent mixing are competitive mechanisms. Fig. 36A and B shows that, as soon as the contribution of the subgrid turbulent mixing appears ($z \simeq 0.4$ m), results from simulation MESHA2 tend to deviate from those of simulation MESHA1. The slopes of the axial profiles of both methane and carbon dioxide become slightly lower, meaning that the overall reactions (R_1 and R_2) become slower. This results in a progressive increase of the CO molar fraction and in a displacement of the combustion zone toward the top. Here, the solid volume fraction is lower and therefore the interphase heat transfer is less effective. As a consequence, an increase of the gas temperature is obtained. In conclusion, combining the chemical reaction rate with the subgrid turbulent mixing rate, in order to model the combustion process, does not improve the numerical predictions. On the one hand, the DG/ED model is useless inside the bed since, in such a zone, the finite-rate chemistry drives the combustion. On the other hand, the DG/ED model decreases the rate of the overall combustion process above the bed, leading to larger amounts of the gaseous reactant species (CH_4 and CO). This effect, combined with a decrease of the solid mass flux, results in an excess of heat in the gas phase, leading to an increase of the maximum of the gas temperature.

6. CONCLUSION

In this study, an Euler–Euler approach was used to perform unsteady 3D numerical simulations of a premixed methane–air combustion, in a dense fluidized bed reactor, containing inert particles. Four different grids were used in order to investigate the influence of the mesh refinement on the combustion. Comparisons between numerical and experimental results pointed out a strong dependency of the gas temperature predictions on the mesh refinement. This dependency was attributed to the hydrodynamic resolution of the bed. Finer meshes make it possible to accurately reproduce the eruption zone of the bed, strongly affecting the interphase heat exchange close to the bed surface and, therefore, the reactive behavior at the selected operating point ($T_{bed} = 700°C$). The latter should indeed correspond to an intermediate regime in which combustion takes place close to the bed surface, because of the ignition of bubbles which are leaving the bed surface (Żukowski, 2003). In this sense, an analysis at mesoscopic and macroscopic scales showed the crucial role of the toroidal loop, which extends from the bed, close to the ejection zone, to the freeboard, in predicting the fuel conversion. A large fuel conversion was indeed found corresponding to the solid recirculation zone. It is believed that this zone should be accurately reproduced in order to provide reliable combustion predictions in dense fluidized beds. The effect of the subgrid turbulent mixing on the combustion was evaluated by using an eddy dissipation model. Preliminary results showed that such a model is not fully appropriate to reproduce the combustion in a dense fluidized bed, at this operating point. The intricate mechanisms which dominate the reactive flow behavior in the splashing zone are not accurately captured by the model. The latter leads to a displacement of the combustion location toward regions where the solid volume fraction is lower and therefore the interphase heat transfer is less effective. The consequence is an overheating of the gas in the freeboard and over. The importance of the recirculation zone and the effectiveness of a combustion model should be further investigated over different regimes in which combustion occurs over the bed or inside. Moreover, the question of the subgrid effects on the hydro-thermodynamics should also be addressed. It is well known that subgrid models make it possible to correctly predict the interaction between the phases so to enable the use of coarser grids (Holloway and Sundaresan, 2012; Özel et al., 2013; Parmentier et al., 2012). Here, it is wondering if appropriate subgrid models would be able to reproduce the effects observed in the eruption zone on the reactive flows. This is left as a future work.

ACKNOWLEDGMENTS

This work received funding from the European Community through the SUCCESS project under the 7th Framework program (Grant agreement No. 608571). It reflects only the author's views and the Community is not liable for any use that may be made of the information contained therein. This work was granted access to the HPC resources of CALMIP supercomputing center under the allocation 2016-0111 and of CINES supercomputing center under the allocation A0022B06012. CALMIP and CINES are gratefully acknowledged. The authors would like to thank Dr. Cyril Baudry and Dr. Jérôme Laviéville for the numerical support with the NEPTUNE_CFD code.

APPENDIX. MODELING REACTIVE GAS–PARTICLE FLOWS

The Eulerian–Eulerian approach relies on the fundamental description of the gaseous and the dispersed phases as interpenetrating continua. Both the phases can be found at the same time in the same computational volume, according to their own volume fractions, α_k. The conservation equations for the gaseous phase are then coupled with that of the dispersed phase through interphase transfer terms. For each phase k, closure models have thus to be provided. The Eulerian–Eulerian approach, used in this work, is based on a statistical description of the dispersed phase using a kinetic particle probability density function (PDF) equation and closure models allowing to account for the interaction with the fluid turbulence and for particle collisions. For details about the approach development, the reader is referred to the work of Simonin (2000). A brief description of the final equations and related closures is given in the following sections.

A.1 Gas-Phase Modeling

The transport equations of the gas phase are derived from the local-instantaneous single-phase equations, accounting for the fluid volume fraction and spatially averaging over a control volume which is much larger than the particle size but smaller than the smallest length scale of the undisturbed flow.[a] The macroscopic conservation equations are thus obtained using a density-weighted averaging (Gobin et al., 2003; Simonin, 2000). Mass, momentum, and enthalpy low Mach equations are written as follow:

$$\frac{\partial}{\partial t}\left(\alpha_g\rho_g\right) + \frac{\partial}{\partial x_j}\left(\alpha_g\rho_g U_{g,j}\right) = \Gamma_g \tag{A.1}$$

$$\alpha_g\rho_g\left(\frac{\partial U_{g,i}}{\partial t} + U_{g,j}\frac{\partial U_{g,i}}{\partial x_j}\right) = -\alpha_g\frac{\partial P_g}{\partial x_i} + \alpha_g\rho_g g_i + \frac{\partial \Sigma_{g,ij}}{\partial x_j} + I_{p\rightarrow g,i} + [U_{\sigma,i} - U_{g,i}]\Gamma_g \tag{A.2}$$

$$\alpha_g\rho_g\left(\frac{\partial H_g}{\partial t} + U_{g,j}\frac{\partial H_g}{\partial x_j}\right) = \frac{\partial}{\partial x_j}\left(\alpha_g\rho_g K_g^{eff}\frac{\partial H_g}{\partial x_j}\right) + \Pi_{p\rightarrow g} + [H_\sigma - H_g]\Gamma_g, \tag{A.3}$$

where the repeated cartesian subscript "j" implies summation. α_g and ρ_g are the gas volume fraction and density, respectively. $U_{g,j} = \langle u_{g,j}\rangle_g$ is the jth component of the mean gas velocity; $\langle \psi_g\rangle_g$ is the density-weighted average of any variable ψ_g. P_g and H_g are the gas pressure and the gas mass enthalpy, respectively, and K_g^{eff} is an effective thermal diffusivity derived using the Boussinesq eddy-viscosity assumption. The term on the right-hand side (r.h.s.) of Eq. (A.1) and the last terms in Eqs. (A.2)

[a] An undisturbed gas flow is the flow that would exist in the absence of a given particle. The single-phase turbulence is then characterized by a range of length and time scales from the Kolmogorov scale to the integral scale.

and (A.3) account, respectively, for the changing of mean mass, momentum and enthalpy due to mass transfers from the particulate phase "p" to the gaseous phase "g", which arise from the gas–solid reactions. In such reactions, the gaseous species cross the interface with the velocity $U_{\sigma,i}$ and enthalpy H_σ and have a total mass transfer rate Γ_g. In this study, no mass transfers between the phases occur, thus $\Gamma_g = 0$. The fourth term $(I_{p\rightarrow g,i})$ in Eq. (A.2) and the second term $(\Pi_{p\rightarrow g})$ in Eq. (A.3) account, respectively, for the interphase momentum and enthalpy exchanges between the phases. These terms will be detailed in Appendix A.3. The third term on the r.h.s. of Eq. (A.2) is the momentum transport due to the velocity fluctuations and viscous stresses. It is written in terms of the effective stress tensor, $\Sigma_{g,ij}$:

$$\Sigma_{g,ij} = -\alpha_g \rho_g R_{g,ij} + \Theta_{g,ij} = -\alpha_g \rho_g \langle u''_{g,i} u''_{g,j} \rangle_g + \Theta_{g,ij}, \tag{A.4}$$

with $u''_{g,i} = u_{g,i} - U_{g,i}$. The quantities $R_{g,ij}$ and $\Theta_{g,ij}$ represent the turbulent-Reynolds and the viscous stress tensors, respectively:

$$\Theta_{g,ij} = \alpha_g \mu_g \left(\frac{\partial U_{g,i}}{\partial x_j} + \frac{\partial U_{g,j}}{\partial x_i} - \frac{2}{3} \frac{\partial U_{g,m}}{\partial x_m} \delta_{ij} \right). \tag{A.5}$$

The turbulent-Reynolds stress tensor stems from the averaging procedure and needs to be modeled or computed from additional transport equations in order to close the momentum transport equation (A.2). In this work, the so-called k-ε eddy-viscosity model is used and complemented by additional terms accounting for the interactions with the dispersed phase (Boëlle et al., 1995; Gobin et al., 2003). The turbulent-Reynolds stress tensor is then written as:

$$\langle u''_{g,i} u''_{g,j} \rangle_g = -\nu_g^t \left(\frac{\partial U_{g,i}}{\partial x_j} + \frac{\partial U_{g,j}}{\partial x_i} \right) + \frac{2}{3} \delta_{ij} \left(k + \nu_g^t \frac{\partial U_{g,m}}{\partial x_m} \right). \tag{A.6}$$

In this equation, ν_g^t is the turbulent viscosity which is modeled as follows (Vermorel et al., 2003):

$$\nu_g^t = \frac{2}{3} k \tau_g^t \left[1 + C_{12} \frac{\alpha_p \rho_p}{\alpha_g \rho_g} \frac{\tau_{gp}^t}{\tau_{gp}^F} \left(1 - \frac{q_{gp}}{2k} \right) \right]^{-1}, \tag{A.7}$$

with $\tau_g^t = C_\mu \frac{3}{2} \frac{k}{\varepsilon}$ the characteristic timescale of the turbulence. α_p and ρ_p are the solid volume fraction and density, respectively. τ_{gp}^t and τ_{gp}^F are timescales related to the interaction between the continuous and the dispersed phases and q_{gp} is the fluid–particle velocity covariance. These quantities will be presented together with the particulate phase closures in Appendix A.3.

In order to be closed, Eqs. (A.6) and (A.7) require the solution of the following transport equations for the gas turbulent kinetic energy, k, and the dissipation rate, ε, as well:

$$\alpha_g\rho_g\left(\frac{\partial k}{\partial t}+U_{g,j}\frac{\partial k}{\partial x_j}\right)=\underbrace{\frac{\partial}{\partial x_j}\left(\alpha_g\rho_g\frac{\nu_g^t}{\sigma_k}\frac{\partial k}{\partial x_j}\right)}_{\text{Diffusion}}\underbrace{-\alpha_g\rho_g\langle u''_{g,i}u''_{g,j}\rangle_g\frac{\partial U_{g,i}}{\partial x_j}}_{\text{Production}}-\underbrace{\alpha_g\rho_g\varepsilon}_{\text{Dissipation}}+\underbrace{\Pi^k_{p\to g}}_{\text{Two-way coupling}} \tag{A.8}$$

$$\alpha_g\rho_g\left(\frac{\partial \varepsilon}{\partial t}+U_{g,j}\frac{\partial \varepsilon}{\partial x_j}\right)=\underbrace{\frac{\partial}{\partial x_j}\left(\alpha_g\rho_g\frac{\nu_g^t}{\sigma_\varepsilon}\frac{\partial \varepsilon}{\partial x_j}\right)}_{\text{Diffusion}}\underbrace{-\alpha_g\rho_g\frac{\varepsilon}{k}C_{\varepsilon_1}\langle u''_{g,i}u''_{g,j}\rangle_g\frac{\partial U_{g,i}}{\partial x_j}}_{\text{Production}}\underbrace{-\alpha_g\rho_g C_{\varepsilon_2}\frac{\varepsilon^2}{k}}_{\text{Dissipation}}+\underbrace{\Pi^\varepsilon_{p\to g}}_{\text{Two-way coupling}} \tag{A.9}$$

The transport equations for k and ε make appear two specific terms accounting for the effect of the particles on the gas, written assuming small particle size with respect to fluid turbulent energetic scales (two-way coupling):

$$\Pi^k_{p\to g}=\frac{\alpha_p\rho_p}{\tau^F_{gp}}\left(-2k+q_{gp}+V_{d,i}V_{r,i}\right),\quad \Pi^\varepsilon_{p\to g}=C_{\varepsilon_3}\frac{\varepsilon}{k}\Pi^k_{p\to g}. \tag{A.10}$$

Table A.1 Constants for the Gas-Phase Turbulence Modeling

C_{12}	C_μ	σ_k	σ_ε	C_{ε_1}	C_{ε_2}	C_{ε_3}
0.34	0.09	1	1.3	1.44	1.92	1.2

The mean velocities $V_{r,i}$ and $V_{d,i}$ will be defined later (Appendix A.3). Constants required for the turbulence modeling are summarized in Table A.1.

A.2 Dispersed Phase Modeling

Eulerian transport equations for the dispersed phase are obtained using the fluid–particle joint pdf approach (Reeks et al., 2015; Simonin, 2000). The final system of equations may be written as follows:

$$\frac{\partial n_p}{\partial t} + \frac{\partial}{\partial x_j}\left[n_p U_{p,j}\right] = C(1) + \frac{\partial}{\partial x_j}\left[-n_p\{u''_{p,j}\}_p\right] + \gamma_p \tag{A.11}$$

$$\frac{\partial}{\partial t}\left[\alpha_p\rho_p\right] + \frac{\partial}{\partial x_j}\left[\alpha_p\rho_p U_{p,j}\right] = \Gamma_p \tag{A.12}$$

$$\begin{aligned}\frac{\partial}{\partial t}\left[\alpha_p\rho_p U_{p,i}\right] + \frac{\partial}{\partial x_j}\left[\alpha_p\rho_p U_{p,i}U_{p,j}\right] &= C(m_p u_{p,i}) + \frac{\partial}{\partial x_j}\left[-\alpha_p\rho_p\langle u''_{p,i}u''_{p,j}\rangle_p\right] \\ &\quad - \alpha_p\frac{\partial P_g}{\partial x_i} + \alpha_p\rho_p g_i + \alpha_p\rho_p\left\langle\frac{F_{r,i}}{m_p}\right\rangle_p \\ &\quad + U_{\sigma,i}\Gamma_p\end{aligned} \tag{A.13}$$

$$\begin{aligned}\frac{\partial}{\partial t}\left[\alpha_p\rho_p H_p\right] + \frac{\partial}{\partial x_j}\left[\alpha_p\rho_p U_{p,j}H_p\right] &= C(m_p h_p) + \frac{\partial}{\partial x_j}\left[-\alpha_p\rho_p\langle u''_{p,i}h''_p\rangle_p\right] \\ &\quad + \alpha_p\rho_p\left\langle\frac{S_{ext}}{m_p}\right\rangle_p + \alpha_p\rho_p\left\langle\frac{Q_r}{m_p}\right\rangle_p \\ &\quad + H_\sigma\Gamma_p\end{aligned} \tag{A.14}$$

In the frame of the Eulerian approach, the solid volume fraction is defined as $\alpha_p = n_p\{m_p\}_p/\rho_p$, where n_p is the mean number of particles per unit volume and $\{m_p\}_p$ the mean solid mass. According to such an approach, any variable ψ_p is composed of its mean $\langle\psi_p\rangle_p$ and fluctuating ψ''_p contributions: $\psi_p = \langle\psi_p\rangle_p + \psi''_p$. The velocity and the mass enthalpy of the dispersed phase are therefore defined as $U_{p,i} = \langle u_{p,i}\rangle_p$ and $H_p = \langle h_p\rangle_p$. The operators $\{\cdot\}_p$ and $\langle\cdot\rangle_p$ represent, respectively, the average and the mass-weighted average over the particle property space (Simonin, 2000). In the above equations, the first r.h.s. term ($C(m_p\psi_p)$) is the mean collisional rate of change of the transported particle property and represents the integral of all possible collisions of the change in $(m_p\psi_p)$. In Eqs. (A.11) and (A.15), this term is taken into account if the particles break up or coalesce during the collisions. In Eq. (A.13), it is found negligible in dilute flows ($\alpha_p < 1\%$), compared to the kinetic stress contribution ($\langle u''_{p,i}u''_{p,j}\rangle_p$), but it should be taken into account in dense flows since the interparticle collisions become important in this regime. In Eq. (A.14), it accounts for the heat exchanged by contact during the interparticle collision; its contribution is negligible, $C(m_p h_p) \simeq 0$. The second term in Eq. (A.11) represents the change in particle number density due to turbulent mixing of particles with different diameter or density. In Eqs. (A.13) and (A.14), it represents, respectively, the momentum and the enthalpy transport by the velocity fluctuations. These terms are either modeled using the Boussinesq approximation or computed from an additional transport equations (Simonin, 2000). The last terms in Eqs. (A.12)–(A.14) account for the influence of the interphase mass transfer; these are null quantities in the present work.

Instead of Eq. (A.11), it is possible to write a transport equation for the mean number of particles per unit mass of the dispersed phase, X_d, defined as $\mathrm{X}_d = 1/\{m_p\}_p$:

$$\frac{\partial}{\partial t}\left[\alpha_p\rho_p\mathrm{X}_d\right] + \frac{\partial}{\partial x_j}\left[\alpha_p\rho_p U_{p,j}\mathrm{X}_d\right] = C(1) + \frac{\partial}{\partial x_j}\left[-\alpha_p\rho_p\langle \mathrm{x}''_d u''_{p,j}\rangle_p\right] + \gamma_p. \quad \text{(A.15)}$$

The last term in Eqs. (A.11) and (A.15) represents the rate of particle disappearance defined as $\gamma_p = n_p\{\frac{dm_p}{dt}\delta(m_p)\}_p$. In this study $\gamma_p = 0$.

A.3 Closure Laws

A.3.1 Modeling the Collision Terms

The integral collision term $C(m_p\psi_p)$ is decomposed in two contributions (Dahler and Sather, 1963 and Jenkins and Richman, 1986), a collisional source term (χ_p) and a collisional flux term ($\Theta_{p,j}$):

$$C(m_p\psi_p) = \chi_p(m_p\psi_p) + \frac{\partial}{\partial x_j}\Theta_{p,j}(m_p\psi_p). \quad \text{(A.16)}$$

In this study, it is assumed that collisions occur without breakage and agglomeration. Therefore, the number of the particles does not change during collisions and $C(1) = 0$ in Eqs. (A.11) and (A.15). For the momentum transport equation, Eq. (A.16) is used to model the collision term $C(m_p u_{p,i})$. Since the collisional source term, $\chi_p(m_p u_{p,i})$, is null, then only the collisional flux term (collisional stress), $\Theta_{p,j}(m_p u_{p,i})$, has to be modeled. The latter is considered as a part of the effective particle stress tensor, $\Sigma_{p,ij}$, for which the modeling is detailed later (see Appendix A.3.3).

A.3.2 Modeling the Mean Interphase Transfers

The mean interphase gas-to-particle momentum transfer

$$I_{g\to p,i} = \alpha_p \rho_p \left\langle \frac{F_{r,i}}{m_p} \right\rangle_p \tag{A.17}$$

satisfies the condition $I_{g\to p,i} + I_{p\to g,i} = 0$. It is modeled assuming that the particles are hard spheres and move by translation in a nonuniform flow:

$$I_{g\to p,i} = -\frac{\alpha_p \rho_p}{\tau_{gp}^F} V_{r,i}. \tag{A.18}$$

$V_{r,i}$ is the mean relative velocity, written in terms of mean phase velocities ($U_{g,i}$ and $U_{p,i}$) and fluid–particle turbulence drift velocity ($V_{d,i}$) due to the turbulent correlation between the instantaneous particle distribution and the fluid velocity field:

$$V_{r,i} = (U_{p,i} - U_{g,i}) - V_{d,i}. \tag{A.19}$$

In the frame of the macroscopic approach, the drift velocity accounts for the turbulence transport of the dispersed phase by the fluid turbulence and is modeled using the following gradient approximation (Fotovat et al., 2015):

$$V_{d,i} = -D_{gp}^t \left(\frac{1}{\alpha_p} \frac{\partial \alpha_p}{\partial x_i} - \frac{1}{\alpha_g} \frac{\partial \alpha_g}{\partial x_i} \right). \tag{A.20}$$

The dispersion coefficient, D_{gp}^t, is proportional to the fluid–particle velocity covariance, q_{gp} and to an eddy-particle interaction time, τ_{gp}^t, characterizing the fluid turbulence seen (*viewed*) by the particles. Its expression is given in Table A.2.

For an ensemble of particles, modeled as a dispersed phase interacting with the fluid turbulence, a mean particle dynamic relaxation time, τ_{gp}^F, is defined. The model considered in this study uses two different correlation laws depending on the value of α_p, Wen and Yu and Ergun laws (Gobin et al., 2003; Simonin et al., 2016):

$$\frac{1}{\tau_{gp}^F} = \frac{3\rho_g}{4\rho_p} \frac{\langle |\mathbf{v_r}| \rangle_p}{d_p} C_D(\langle \mathrm{Re} \rangle_p) \tag{A.21}$$

Table A.2 Closure Parameters for the $q_p^2 - q_{gp}$ Model

$\Lambda_p = \alpha_p \rho_p \frac{4}{3} \alpha_p g_0 (1+e_c) d_p \sqrt{\frac{2}{3}\frac{q_p^2}{\pi}}$		
$\Phi_c = \frac{2}{5}(1+e_c)(3e_c - 1)$	$\sigma_c = \frac{(1+e_c)(3-e_c)}{5}$	
$\varphi_c = \frac{3}{5}(1+e_c)^2(2e_c - 1)$	$\xi_c = \frac{(1+e_c)(49-33e_c)}{100}$	
$g_0 = \left(1 - \frac{\alpha_p}{\alpha_m}\right)^{-2.5\alpha_m}$ with $\alpha_m = 0.64$		
$D_{gp}^t = \tau_{gp}^t \frac{1}{3} q_{gp}$	$\nu_{gp}^t = \frac{1}{3}\tau_{gp}^t q_{gp}$	$\eta_r = \frac{\tau_{gp}^t}{\tau_{gp}^F}$

$$\text{with} \quad C_D(\langle \mathrm{Re} \rangle_p) = \begin{cases} C_{D,WY} & \text{if } \alpha_g \geq 0.7, \\ \min\left[C_{D,WY}; C_{D,Erg}\right] & \text{if } \alpha_g < 0.7. \end{cases} \tag{A.22}$$

$$C_{D,WY} = \begin{cases} \frac{24}{\langle \mathrm{Re} \rangle_p}\left[1 + 0.15 \langle \mathrm{Re} \rangle_p^{0.687}\right] \alpha_g^{-1.7} & \text{for } \langle \mathrm{Re} \rangle_p < 1000 \\ 0.44 \alpha_g^{-1.7} & \text{for } \langle \mathrm{Re} \rangle_p \geq 1000 \end{cases} \tag{A.23}$$

$$C_{D,Erg} = 200 \frac{1 - \alpha_g}{\langle \mathrm{Re} \rangle_p} + \frac{7}{3} \quad \text{and} \quad \langle \mathrm{Re} \rangle_p = \frac{\alpha_g \langle |\mathbf{v_r}| \rangle_p d_p}{\nu_g}. \tag{A.24}$$

The mean interphase enthalpy transfer is defined as:

$$\Pi_{g \to p} = \alpha_p \rho_p \left\langle \frac{Q_r}{m_p} \right\rangle_p \tag{A.25}$$

and satisfies the condition $\Pi_{g \to p} + \Pi_{p \to g} = 0$. It can be modeled as follows:

$$\Pi_{g \to p} = -\frac{\alpha_p \rho_p C_{P_p}}{\tau_{gp}^T} T_r, \tag{A.26}$$

where T_r is the mean relative temperature, $T_r \simeq T_p - T_g$. The thermal relaxation time, τ_{gp}^T, is defined as

$$\frac{1}{\tau_{gp}^T} = \frac{6\lambda_g}{\rho_p C_{P_p}} \frac{\langle \mathrm{Nu} \rangle_p}{d_p^2}, \tag{A.27}$$

based on the mean Nusselt number, $\langle \text{Nu} \rangle_p$, given in terms of Reynolds and Prandtl numbers, Re and Pr. In this study, the correlation proposed by Sun et al. (2015) was used:

$$\begin{aligned}\langle \text{Nu} \rangle_p &= \left(-0.46 + 1.77\alpha_g + 0.69\alpha_g^2\right)/\alpha_g^3 \\ &+ \left(1.37 - 2.4\alpha_g + 1.2\alpha_g^2\right)\langle \text{Re} \rangle_p^{0.7} \text{Pr}^{1.3}.\end{aligned} \tag{A.28}$$

A.3.3 Modeling the Particle Fluctuating Motion

The effective particle stress tensor, $\Sigma_{p,ij}$, is written as:

$$\Sigma_{p,ij} = -\alpha_p \rho_p R_{p,ij} + \Theta_{p,ij} = -\alpha_p \rho_p \langle u''_{p,i} u''_{p,j} \rangle_p + \Theta_{p,ij}. \tag{A.29}$$

The kinetic part of the effective particle stress tensor is found to be dominant in dilute flows, whereas the collisional part is dominant in dense flows. The modeling of those terms is derived in the frame of the KTGF (Jenkins and Richman, 1986), accounting for the effect of the interstitial fluid (Simonin, 2000). So, both the kinetic and the collisional parts of the effective stress will be functions of the particle agitation and the fluid–particle correlation.

The particle kinetic stress tensor is modeled using the Boussinesq approximation (Boëlle et al., 1995; Gobin et al., 2003; Simonin, 2000):

$$\langle u''_{p,i} u''_{p,j} \rangle_p = -\nu_p^{kin} \left(\frac{\partial U_{p,i}}{\partial x_j} + \frac{\partial U_{p,j}}{\partial x_i} \right) + \frac{2}{3}\delta_{ij} \left(q_p^2 + \nu_p^{kin} \frac{\partial U_{p,m}}{\partial x_m} \right). \tag{A.30}$$

The particle kinetic viscosity, ν_p^{kin}, accounts directly for the combined effects of different mechanisms, such as the transport of the particle momentum by the fluid turbulence and by the random motion of the particles:

$$\nu_p^{kin} = \left[\nu_{gp}^t + \frac{\tau_{gp}^F}{2} \frac{2}{3} q_p^2 \left(1 + \alpha_p g_0 \Phi_c\right) \right] \left(1 + \frac{\tau_{gp}^F}{2} \frac{\sigma_c}{\tau_p^c} \right)^{-1}. \tag{A.31}$$

According to Boëlle et al. (1995), the collisional part of the effective particle stress tensor is not affected by the interstitial fluid and therefore the corresponding Boussinesq approximation is nearly identical to the one developed for dry granular media (Jenkins and Richman, 1986):

$$\begin{aligned}\Theta_{p,ij} &= -\left[\frac{2}{3}\alpha_p \rho_p q_p^2 2\alpha_p g_0 (1 + e_c) - \Lambda_p \frac{\partial U_{p,m}}{\partial x_m} \right] \delta_{ij} \\ &+ \alpha_p \rho_p \nu_p^{col} \left(\frac{\partial U_{p,i}}{\partial x_j} + \frac{\partial U_{p,j}}{\partial x_i} - \frac{2}{3} \frac{\partial U_{p,m}}{\partial x_m} \delta_{ij} \right),\end{aligned} \tag{A.32}$$

with the collisional viscosity

$$\nu_p^{col} = \frac{4}{5}\alpha_p g_0 (1+e_c)\left(\nu_p^{kin} + d_p\sqrt{\frac{2}{3}\frac{q_p^2}{\pi}}\right). \tag{A.33}$$

The parameter e_c is the restitution coefficient and is defined as the ratio of the relative radial velocities of a colliding pair of particles after and prior to impact. It accounts for particle energy exchanges in partially elastic collisions. The function g_0 is the radial distribution function and, in the Enskog kinetic theory of gases, it takes into account the increased probability of collisions in dense gases compared to dilute gases, for which the function g_0 is equal to unity. A transport equation is used for the particle fluctuant kinetic energy, $q_p^2 = \langle u''_{p,i} u''_{p,i}\rangle_p/2$. It is obtained by adding the separate transport equations governing the normal kinetic stress tensor components (see reference Simonin, 2000):

$$\alpha_p\rho_p\left(\frac{\partial q_p^2}{\partial t} + U_{p,j}\frac{\partial q_p^2}{\partial x_j}\right) = \frac{\partial}{\partial x_j}\left(\alpha_p\rho_p\kappa_p^{eff}\frac{\partial q_p^2}{\partial x_j}\right) + \Sigma_{p,ij}\frac{\partial U_{p,i}}{\partial x_j} - \alpha_p\rho_p\varepsilon_p + \Pi_{q_p}. \tag{A.34}$$

The turbulent diffusivity coefficient is defined as $\kappa_p^{eff} = \kappa_p^{kin} + \kappa_p^{col}$, where

$$\kappa_p^{kin} = \left[\frac{1}{3}\tau_{gp}^t q_{gp} + \frac{5}{9}\tau_{gp}^F\frac{2}{3}q_p^2(1+\alpha_p g_0\varphi_c)\right]\left(1+\frac{5}{9}\tau_{gp}^F\frac{\xi_c}{\tau_p^c}\right)^{-1}, \tag{A.35}$$

$$\kappa_p^{col} = \alpha_p g_0(1+e_c)\left[\frac{6}{5}\kappa_p^{kin} + \frac{4}{3}d_p\sqrt{\frac{2}{3}\frac{q_p^2}{\pi}}\right]. \tag{A.36}$$

In Eq. (A.34), the quantity ε_p is the particle kinetic energy dissipation rate due to inelastic collisions and is written as:

$$\varepsilon_p = \frac{1}{3}(1-e_c^2)\frac{q_p^2}{\tau_p^c}. \tag{A.37}$$

The term Π_{q_p} is the interphase turbulent kinetic energy transfer rate

$$\Pi_{q_p} = -\alpha_p\rho_p\frac{1}{\tau_{gp}^F}(2q_p^2 - q_{gp}). \tag{A.38}$$

The change of the mean number of particles per unit volume or mass (Eqs. (A.11) and (A.15)) due to turbulent mixing of particles with different diameter or density ($n_p\{u''_{p,j}\}_p = \alpha_p\rho_p\langle u''_{p,j}/m_p\rangle_p = \alpha_p\rho_p\langle \mathrm{x}''_d u''_{p,j}\rangle_p$) is modeled using the following approximation (Laviéville et al., 1995)

$$\langle \mathrm{x}''_d u''_{p,j}\rangle_p = -D_p^{kin}\nabla \mathrm{X}_d, \tag{A.39}$$

with

$$D_p^{kin} = \left[D_{gp}^t + \tau_{gp}^F \frac{2}{3} q_p^2\right]\left(1 + \frac{2}{3}\frac{\tau_{gp}^F}{\tau_p^c}\right)^{-1}. \tag{A.40}$$

The term accounting for the transport of the enthalpy by the velocity fluctuations in Eq. (A.14) is also modeled using a Boussinesq approximation as follows:

$$\langle h''_p u''_{p,j}\rangle_p = -K_p^{kin}\nabla H_p. \tag{A.41}$$

where $K_p^{kin} = D_p^{kin}$. The different parameters appearing in these equations are given in Table A.2, with the collision rate based on the modeling accounting for the effect of particle–particle velocity correlation on the collisions (Simonin et al., 2002). A summary of the different timescales (including τ_{gp}^t and τ_p^c) used in the modeling is then given in Table A.3.

A.3.4 Modeling the Fluid–Particle Velocity Correlations

The transport equation for q_{gp}, with $q_{gp} = \langle u''_{g,i}u''_{p,i}\rangle_p$, is derived in the frame of the fluid–particle joint pdf approach from the separate transport equations governing the fluid–particle correlation tensor components $\langle u''_{g,i}u''_{p,j}\rangle_p$ (Février and Simonin, 1998; Simonin et al., 1993):

$$\alpha_p\rho_p\left(\frac{\partial q_{gp}}{\partial t} + U_{p,j}\frac{\partial q_{gp}}{\partial x_j}\right) = \frac{\partial}{\partial x_j}\left(\alpha_p\rho_p\frac{\nu_{gp}^t}{\sigma_k}\frac{\partial q_{gp}}{\partial x_j}\right) - \alpha_p\rho_p\varepsilon_{gp} + \Pi_{q_{gp}} - \alpha_p\rho_p\left[\langle u''_{g,i}u''_{p,j}\rangle_p\frac{\partial U_{p,i}}{\partial x_j} + \langle u''_{g,j}u''_{p,i}\rangle_p\frac{\partial U_{g,i}}{\partial x_j}\right], \tag{A.42}$$

Table A.3 Characteristic Timescales Used in the Eulerian Approach

Mean particle dynamic relaxation time:

It represents the fluid–particle interaction through the drag force

$$\frac{1}{\tau_{gp}^{F}} = \frac{3\rho_g}{4\rho_p}\frac{\langle|\mathbf{v_r}|\rangle_p}{d_p}\langle C_D\rangle$$

Fluid turbulent timescale:

It is related to τ_p through the Stokes' number, $\mathrm{St} = \tau_p/\tau_g^t$

$$\tau_g^t = C_\mu \frac{3}{2}\frac{k}{\varepsilon}$$

Eddy-particle interaction time:

It is the characteristic time from the turbulence seen by the particles

$$\tau_{gp}^{t} = \frac{\tau_g^t}{\sigma_k}\left(1 + C_\beta \frac{V_{r,i}V_{r,i}}{\frac{2}{3}k}\right)^{-1/2}$$

Interparticle collision time:

It corresponds to the mean collision time for mono-dispersed phase

– *Assumption of uncorrelated collisions:*

$$\tau_{p,\,uncorr}^{c} = \left(6\frac{\alpha_p g_0}{d_p}\sqrt{\frac{16}{\pi}\frac{2}{3}q_p^2}\right)^{-1}$$

– *Assumption of correlated collisions* (Simonin et al., 2002)

Account for correlation between colliding particles due to the interaction with the fluid turbulence

$$\tau_p^c = \tau_{p,\,uncorr}^{c}\left[1 - \zeta_{gp}^2\right]^{-1/2} \text{ with } \zeta_{gp}^2 = \frac{[q_{qp}]^2}{4kq_p^2},\ 0 < \zeta_{gp}^2 < 1$$

Mean particle thermal relaxation time:

It accounts for the interphase heat exchange

$$\frac{1}{\tau_{gp}^{T}} = \frac{6\lambda_g}{\rho_p C_{P_p}}\frac{\langle \mathrm{Nu}\rangle_p}{d_p^2}$$

where ε_{gp} is the fluid–particle covariance dissipation rate due to viscous dissipation and crossing trajectory effects:

$$\varepsilon_{gp} = \frac{q_{gp}}{\tau_{gp}^{t}}. \tag{A.43}$$

The interphase interaction term is written as:

$$\Pi_{q_{gp}} = -\alpha_p\rho_p\frac{1}{\tau_{gp}^{F}}\left[(q_{gp} - 2k) + \frac{\alpha_p\rho_p}{\alpha_g\rho_g}(q_{gp} - 2q_p^2)\right]. \tag{A.44}$$

The fluid–particle velocity correlation tensor $\langle u''_{g,i} u''_{p,j} \rangle_p$ is modeled using the Boussinesq approximation (Février and Simonin, 1998; Gobin et al., 2003; Simonin, 2000):

$$\begin{aligned}\langle u''_{g,i} u''_{p,j} \rangle_p = & \frac{1}{3} q_{gp} \delta_{ij} + \frac{\eta_r}{1+\eta_r} \left[\langle u''_{g,i} u''_{g,j} \rangle_g - \frac{2}{3} k \delta_{ij} \right] \\ & - \frac{\nu^t_{gp}}{1+\eta_r} \left[\frac{\partial U_{g,i}}{\partial x_j} + \frac{\partial U_{p,j}}{\partial x_i} - \frac{1}{3} \delta_{ij} \left(\frac{\partial U_{g,m}}{\partial x_m} + \frac{\partial U_{p,m}}{\partial x_m} \right) \right].\end{aligned} \tag{A.45}$$

REFERENCES

Alobaid F, Ohlemüller P, Ströhle J, Epple B: Extended Euler-Euler model for the simulation of a 1 MWth chemical-looping pilot plant, *Energy* 93(Pt. 2):2395–2405, 2015.

Ansart R, Garcìa-Triñanes P, Boissière B, Benoit H, Seville JPK, Simonin O: Dense gas-particle suspension upward flow used as heat transfer fluid in solar receiver: PEPT experiments and 3D numerical simulations, *Powder Technol* 307:25–36, 2017.

Anthony E: Fluidized bed combustion of alternative solid fuels; status, successes and problems of the technology, *Prog Energy Combust Sci* 21(3):239–268, 1995.

Barlow RS, Karpetis AN, Frank JH, Chen JY: Scalar profiles and NO formation in laminar opposed-flow partially premixed methane/air flames, *Combust Flame* 127(3):2102–2118, 2001.

Baron J, Bulewicz E, Żukowski W, Kandefer S, Pilawska M: Combustion of hydrocarbon fuels in a bubbling fluidized bed, *Combust Flame* 128(4):410–421, 2002.

Baron J, Bulewicz EM, Zabagło J, Żukowski W: Propagation of reaction between bubbles with a gas burning in a fluidised bed, *Flow Turbul Combust* 88(4):479–502, 2012.

Boëlle A, Balzer G, Simonin O: Second-order prediction of the particle-phase stress tensor of inelastic spheres in simple shear dense suspensions. In *Gas-solid flows, ASME FED*, vol. 228, 1995, ASME, pp 9–18.

Dahler JS, Sather NF: Transport properties of polyatomic fluids, a dilute gas of perfectly rough spheres, *J Chem Phys* 38:2363–2382, 1963.

Dounit S: Combustion du gaz naturel en reacteur à lit fluidisé: Etude expérimentale et modélisation de la zone dense et de la zone de désengagement (Ph.D. thesis), France, 2001, Intitut National Polytechnique de Toulouse.

Dounit S, Hemati M, Steinmetz D: Natural gas combustion in fluidised bed reactors between 600 and 850°C: experimental study and modelling of the freeboard, *Powder Technol* 120(1):49–54, 2001.

Dounit S, Hemati M, Andreux R: Modelling and experimental validation of a fluidized-bed reactor freeboard region: application to natural gas combustion, *Chem Eng J* 140(1):457–465, 2008.

Dryer FL, Glassman I: High-temperature oxidation of CO and CH_4, *Symp Combust* 14(1):987–1003, 1973.

Dufresne Y, Moureau V, Masi E, Simonin O, Horwitz J: Simulation of a reactive fluidized bed reactor using CFD/DEM. In *Proceedings of the summer program 2016,* 2016, Center for Turbulence Research.

Fede P, Simonin O, Ghouila I: 3D numerical simulation of polydisperse pressurized gas-solid fluidized bed. In *Proceedings of AJK2011-FED, ASME-JSME-KSME joint fluids engineering conference 2011, Hamamatsu, Shizuoka (Japan), paper no. AJK2011-12016*, 2011.

Fede P, Simonin O, Ingram A: 3D numerical simulation of a lab-scale pressurized dense fluidized bed focusing on the effect of the particle-particle restitution coefficient and particle-wall boundary conditions, *Chem Eng Sci* 142:215–235, 2016.

Février P, Simonin O: Constitutive relations for fluid-particle velocity correlations in gas-solid turbulent flows. In *Proc. 3rd int. conference on multiphase flow, ICMF'98, Lyon (France)*, 1998.

Foka M, Chaouki J, Guy C, Klvana D: Natural gas combustion in a catalytic turbulent fluidized bed, *Chem Eng Sci* 49(24, Pt. A):4269–4276, 1994.

Fotovat F, Ansart R, Hemati M, Simonin O, Chaouki J: Sand-assisted fluidization of large cylindrical and spherical biomass particles: experiments and simulation, *Chem Eng Sci* 126:543–559, 2015.

Gobin A, Neau H, Simonin O, Llinas JR, Reiling V, Sélo JL: Fluid dynamic numerical simulation of a gas phase polymerization reactor, *Int J Numer Methods Fluids* 43(10–11):1199–1220, 2003.

Gómez-Barea A, Leckner B: Modeling of biomass gasification in fluidized bed, *Prog Energy Combust Sci* 36(4):444–509, 2010.

Grosshandler WL: *RADCAL: a narrow band model for radiation calculations in a combustion environment.* NIST technical note 1402, 1993.

Hamidouche Z: *Modeling and numerical simulation of coupled reactive fluidized beds in a chemical looping combustion system,* Ph.D. Thesis, 2017, *France,* Institut National Polytechnique de Toulouse. http://oatao.univ-toulouse.fr/.

Hamidouche Z, Masi E, Fede P, Simonin O, Ansart R, Hemati M: Unsteady three-dimensional numerical simulations of methane combustion in dense fluidized bed. In *Proceedings of the 9th international conference on multiphase flow, Firenze, Italy, May 22–27*, 2016.

Hesketh R, Davidson JF: Combustion of methane and propane in an incipiently fluidized bed, *Combust Flame* 85:449–467, 1991.

Holloway W, Sundaresan S: Filtered models for reacting gasparticle flows, *Chem Eng Sci* 82:132–143, 2012.

Iamarino M, Ammendola P, Chirone R, Pirone R, Ruoppolo G, Russo G: Nonpremixed catalytic combustion of methane in a fluidized bed reactor, *Ind Eng Chem Res* 45:1009–1013, 2006.

Jenkins JT, Richman MW: Grads 13-moment system for a dense gas of inelastic spheres. In *The breadth and depth of continuum mechanics*, 1986, Springer, pp 647–669.

Konan NA, Neau H, Simonin O, Dupoizat M, Le Goaziou T: Reactive multiphase flow simulation of uranium hexafluoride conversion reactor. In *Proceedings of the 7th international conference on multiphase flow, Tampa, FL, May 30–June 4*, 2010a.

Konan NA, Neau H, Simonin O, Dupoizat M, Le Goaziou T: 3D unsteady multiphase simulation of uranium tetrafluoride particle fluorination in fluidized bed pilot. In Yue G, Zhang H, Zhao C, Luo Z, editors: *Proceedings of the 20th international conference on fluidized bed combustion*, Berlin, Heidelberg, 2010b, Springer, pp 1152–1158.

Kunii D, Levenspiel O: Entrainment of solids from fluidized beds I. Hold-up of solids in the freeboard II. Operation of fast fluidized beds, *Powder Technol* 61(2):193–206, 1990.

Laviéville J, Deutsch E, Simonin O: Large Eddy simulation of interactions between colliding particles and a homogeneous isotropic turbulence field. In *Gas-particle flows, ASME FED*, 1995, vol. 228, pp 347–357.

Laviolette JP, Patience GS, Marca CL, Chaouki J: Gas-phase propane combustion in the freeboard of a fluidized bed, *Fuel* 111:316–323, 2013.

Linstrom PJ, Mallard WG: The NIST Chemistry WebBook: a chemical data resource on the Internet, *J Chem Eng Data* 46(5):1059–1063, 2001.

Loha C, Chattopadhyay H, Chatterjee PK: Three dimensional kinetic modeling of fluidized bed biomass gasification, *Chem Eng Sci* 109:53–64, 2014.

Magnussen BF: On the structure of turbulence and a generalized eddy dissipation concept for chemical reaction in turbulent flow. In *19th AIAA aerospace science meeting, St Louis, USA*, vol. 198, 1981, pp 1.

Magnussen BF: The Eddy dissipation concept—a bridge between science and technology. In *ECCOMAS thematic conference on computational combustion*, 2005, pp 21–24.

Magnussen BF, Hjertager BH: On mathematical modeling of turbulent combustion with special emphasis on soot formation and combustion, *Symp Combust* 16(1):719–729, 1977.

Méchitoua N, Boucker M, Laviéville J, Hérard JM, Pigny S, Serre G: An unstructured finite volume solver for two-phase water/vapour flows modeling based on elliptic oriented fractional step method. In *Proceeding of NURETH, international topical meeting on nuclear reactor thermal hydraulics*, 2003.

Neau H, Laviéville J, Simonin O: NEPTUNE CFD high parallel computing performances for particle-laden reactive flows. In *Proceedings of the 7th international conference on multiphase flow, Tampa, FL, May 30-June 4*, 2010.

Ogada T, Werther J: Combustion characteristics of wet sludge in a fluidized bed: release and combustion of the volatiles, *Fuel* 75(5):617–626, 1996.

Özel A, Fede P, Simonin O: Development of filtered Euler-Euler two-phase model for circulating fluidised bed: high resolution simulation, formulation and *a priori* analyses, *Int J Multiphase Flow* 43–63, 2013.

Parker JM: CFD model for the simulation of chemical looping combustion, *Powder Technol* 265:47–53, 2014.

Parmentier JF, Simonin O, Delsart O: A functional subgrid drift velocity model for filtered drag prediction in dense fluidized bed, *AIChE J* 58(4):1084–1098, 2012.

Pre P, Hemati M, Marchand B: Study on natural gas combustion in fluidized beds: modelling and experimental validation, *Chem Eng Sci* 53(16):2871–2883, 1998.

Pröll T, Kolbitsch P, Bolhàr-Nordenkampf J, Hofbauer H: A novel dual circulating fluidized bed system for chemical looping processes, *AIChE J* 55(12):3255–3266, 2009.

Reeks M, Simonin O, Fede P: PDF models for particle transport, mixing and collisions in turbulent gas flows. In Michaelides EE, Crowe CT, Schwarzkopf JD, editors: *Multiphase flow hand book*, 2015, CRC Press, Taylor and Francis Group, pp 144–202.

Rüdisüli M, Schildhauer TJ, Biollaz SMA, van Ommen JR: Scale-up of bubbling fluidized bed reactors—a review, *Powder Technol* 217:21–38, 2012.

Scala F, Salatino P: Modelling fluidized bed combustion of high-volatile solid fuels, *Chem Eng Sci* 57(7):1175–1196, 2002.

Simonin O: Statistical and continuum modelling of turbulent reactive particulate flows. Part 1: Theoretical derivation of dispersed Eulerian modelling from probability density function kinetic equation. In *Theoretical and experimental modeling of particulate flows, Lecture Series*, vol. 6, von Karman Institute for Fluid Dynamics, 2000, Rhode Saint Genèse, Belgium.

Simonin O, Deutsch E, Minier JP: Eulerian prediction of the fluid/particle correlated motion in turbulent two-phase flows, *Appl Sci Res* 51:275–283, 1993.

Simonin O, Février P, Laviéville J: On the spatial distribution of heavy-particle velocities in turbulent flow: from continuous field to particulate chaos, *J Turbul* 3(1):1–40, 2002.

Simonin O, Chevrier S, Audard F, Fede P: Drag force modelling in dilute to dense particle-laden flows with mono-disperse or binary mixture of solid particles. In *Proceedings of the 9th international conference on multiphase flow, Firenze, Italy, May 22–27*, 2016.

Smith GP, Golden DM, Frenklach M, et al: n.d., http://www.me.berkeley.edu/gri-mech/.

Sotudeh-Gharebagh R, Chaouki J: The heterogeneous and homogeneous combustion of methane over inert particles, *Can J Chem Eng* 81(6):1182–1191, 2003.

Sotudeh-Gharebagh R, Mostoufi N: Simulation of a catalytic turbulent fluidized bed reactor using the sequential modular approach, *Fuel Process Technol* 85(2):189–200, 2004.

Sotudeh-Gharebagh R, Chaouki J, Sauriol P: An experimental study of non-premixed combustion in a turbulent fluidized-bed reactor, *Fuel Process Technol* 88(9):847–858, 2007.

Spalding DB: Mixing and chemical reaction in steady confined turbulent flames, *Symp Combust* 13(1):649–657, 1971.

Spalding DB: Development of the eddy-break-up model of turbulent combustion, *Symp Combust* 16(1):1657–1663, 1977.

Sun B, Tenneti S, Subramaniam S: Modeling average gas-solid heat transfer using particle-resolved direct numerical simulation, *Int J Heat Mass Transf* 86:898–913, 2015.

van der Vaart DR: Freeboard ignition of premixed hydrocarbon gas in a fluidized bed, *Fuel* 67(7):1003–1007, 1988.

Vermorel O, Bedat B, Simonin O, Poinsot T: Numerical study and modelling of turbulence modulation in a particle laden slab flow, *J Turbul* 4(25):1–39, 2003.

Xue Q, Fox RO: Multi-fluid CFD modeling of biomass gasification in polydisperse fluidized-bed gasifiers, *Powder Technol* 254:187–198, 2014.

Żukowski W: Acoustic effects during the combustion of gaseous fuels in a bubbling fluidized bed, *Combust Flame* 117(3):629–635, 1999.

Żukowski W: A simple model for explosive combustion of premixed natural gas with air in a bubbling fluidized bed of inert sand, *Combust Flame* 134(4):399–409, 2003.

CHAPTER THREE

Simulation of Turbulent Coalescence and Breakage of Bubbles and Droplets in the Presence of Surfactants, Salts, and Contaminants

Salvatore Falzone, Antonio Buffo, Marco Vanni, Daniele Luca Marchisio[1]
Institute of Chemical Engineering, Politecnico di Torino, Torino, Italy
[1]Corresponding author: e-mail address: daniele.marchisio@polito.it

Contents

1. Introduction 126
2. Phenomenology of Fluid Particle Breakup 128
3. Phenomenology of Fluid Particle Coalescence 131
4. Analysis of Different Models for Fluid Particle Breakage Kernel 136
   4.1 Turbulent Fluctuations Breakage Models 136
   4.2 Models for Other Breakage Mechanisms 155
   4.3 Breakage Kernel in Contaminated Systems 156
   4.4 Comparison Among Different Breakage Kernel Models 157
5. Models for the Daughter Size Distribution Function (DaSD) 159
   5.1 Statistical Models 159
   5.2 Phenomenological Models 160
6. Analysis of Different Models for Fluid Particle Coalescence Kernel 163
   6.1 Models for Collision Frequency $h(\xi_1, \xi_2)$ 164
   6.2 Models for Coalescence Efficiency $\eta(\xi_1, \xi_2)$ 168
   6.3 Comparison Among Different Models for Coalescence Kernels 179
7. Conclusions 182
References 183

Abstract

The accurate prediction of the size distribution in liquid–liquid or gas–liquid turbulent dispersions is of fundamental relevance in many industrial applications. The distribution can be predicted with computational fluid dynamics coupled with population balance models (PBM). PBM needs, in turn, suitable models to account for droplet and bubble coalescence and breakage. In this work a critical analysis of the most commonly

Advances in Chemical Engineering, Volume 52
ISSN 0065-2377
https://doi.org/10.1016/bs.ache.2018.01.002

employed breakage and coalescence kernels is performed. In particular, the presence of contaminants on breakage and coalescence rates is analyzed and the appropriate modifications on the kernels are discussed. Furthermore, the most important models for the daughter size distribution are reviewed. Eventually, a comparison among the most employed breakage and coalescence model was carried out.

1. INTRODUCTION

Mass, heat, and momentum interfacial transport are fundamental phenomena in every industrial multiphase flow, that depend on the contact area between the continuous and disperse phases; therefore the prediction of this property is of fundamental importance in order to design, optimize, and scale-up multiphase processes. The droplet or bubble size distribution (DSD or BSD), or more generally the fluid particle size distribution (PSD), which defines the total specific contact area in a polydisperse multiphase flow, can be predicted by computational fluid dynamics (CFD) coupled with population balance models (PBM) (Chen et al., 2011; Marchisio and Fox, 2013; Wang and Wang, 2007; Yan et al., 2012; Yao et al., 2015). The PBM defines the evolution of the number density function (NDF), n, through the solution of the so-called population balance equation (PBE). By assuming that the NDF depends only on the size of the fluid particle ξ, namely, the equivalent particle diameter, and considering only particle coalescence and breakage, the PBE reads as follows (by omitting the dependences on space and time):

$$\begin{aligned}\frac{\partial n(\xi)}{\partial t} + \frac{\partial}{\partial x}\cdot(\mathbf{u}_{\mathrm{p}}n(\xi)) &= \int_{\xi}^{\infty}\beta(\xi|\xi')b(\xi')n(\xi')\mathrm{d}\xi' - b(\xi)n(\xi)\\ &+\frac{1}{2}\int_0^{\xi} a\left((\xi^3-\xi'^3)^{1/3},\xi'\right)n\left((\xi^3-\xi'^3)^{1/3}\right)n(\xi')\frac{\xi^2}{(\xi^3-\xi'^3)^{2/3}}\mathrm{d}\xi' \\ &-n(\xi)\int_0^{\infty} a(\xi,\xi')n(\xi')\mathrm{d}\xi'\end{aligned} \quad (1)$$

where the first left-hand side term is the accumulation term, the second left-hand side term represents the changes in the NDF due to the particle convection in the physical space. On the right-hand side, the first two terms describe the birth and death of particles due to breakage, while the last two terms represent the birth and death of particles due to coalescence. More details about Eq. (1) can be found in the book of Marchisio and Fox (2013).

In the last decades, many methods have been developed to solve the PBM, such as the classes method (CM) (Bannari et al., 2008), the Monte Carlo method (MCM) (Lin et al., 2002), and the methods of moments, among which particularly important are quadrature-based moment methods (QBMM), such as the quadrature method of moments (QMOM) (Marchisio et al., 2003) and the direct quadrature method of moments (DQMOM) (Marchisio and Fox, 2005). Among these, it also worth citing the adaptive direct quadrature method of moments (ADQMOM) (Su et al., 2008), the conditional quadrature method of moments (CQMOM) (Yuan and Fox, 2011), and the extended quadrature method of moments (EQMOM) (Yuan et al., 2012).

One major issue in the development of a PBM is the modeling of the coalescence and breakage terms of Eq. (1), namely, $b(\xi)$ and $a(\xi, \xi')$, commonly known as breakage and coalescence kernels, respectively, and of the daughter size distribution (DaSD) $\beta(\xi|\xi')$ of Eq. (1). Although the formulation of kernels and DaSDs has been the subject of many studies, the corresponding models are still at an early stage of development.

The standard framework for fluid particle breakage and coalescence in turbulent flows is that of the classical Kolmogorov theory (Coulaloglou and Tavlarides, 1977; Kolmogorov, 1941; Luo and Svendsen, 1996; Prince and Blanch, 1990) and is focused on the inertial subrange of turbulence. More recent attempts tried to extend the analysis to the entire spectrum of turbulence, thus including the energy containing and the dissipation subranges in addition to the internal subrange (e.g., Ghasempour et al., 2014a,b; Han et al., 2015; Solsvik and Jakobsen, 2016). The importance of the intermittent nature of turbulence especially on breakage was pointed out Baldyga and Bourne (1995) and included in the multifractal kernel by Baldyga and Podgórska (1998). It is worth mentioning here that the possible dependency of the kernels on turbulent quantities, such as the turbulent dissipation rate (ϵ) and the turbulent kinetic energy (k), does not imply the use of any specific turbulence modeling framework (RANS, LES, or DNS). Another aspect that needs to be highlighted is the influence of salts, surfactants, and contaminants of various nature (including ultrafine particles) dissolved in the continuous phase. These can affect the interfacial tension, the Hamaker constant value, and the mobility of the fluid particles at the interface due to Marangoni effects.

The influence of surfactants on fluid particle breakage is usually taken into account only through the reduction of interfacial tension and therefore the decrease of shape-restoring stresses. However, as shown by

Koshy et al. (1988), the maximum stable fluid particle size can be significantly overpredicted if only the steady-state (or static) interfacial tension is considered. It was, in fact, claimed that the difference between dynamic and static interfacial tension can aid turbulent stresses to break the drops. Based on these results, Bak and Podgórska (2012) extended the multifractal kernel to include this mechanism.

Coalescence is also highly sensitive to impurities, even in traces. Surfactants dissolved in the continuous phase decrease the interfacial tension, immobilize fluid interfaces, and slow down the overall coalescence phenomena. However, the importance of surfactants on fluid particle coalescence decreases with the increase in disperse phase viscosity. Moreover, depending on interface mobility and deformation in the contact area, coalescence is favored by energetic or gentle collisions (Chesters, 1991).

The aim of this review is to critically analyse the most popular approaches to describe coalescence and breakage in the turbulent system. In particular, unlike previous works (Lasheras et al., 2002; Liao and Lucas, 2009, 2010; Sajjadi et al., 2013; Solsvik et al., 2013), this review also focuses on how to modify these models in order to take into account the presence of contaminants in real systems.

2. PHENOMENOLOGY OF FLUID PARTICLE BREAKUP

The breakup of fluid particles (droplets and bubbles) in a turbulent flow is considered in PBM a "discrete" process, that depends on the hydrodynamics of the continuous phase and the interactions between the two phases at the interface. A fluid particle immersed in a turbulent flow undergoes turbulent stresses, viscous stresses, and dynamic pressure that act in such a way as to deform and disrupt the particle. The particle may change its shape with the fluctuations of the surrounding fluid or due to collisions with eddies. When the amplitude of the oscillation is close to that required to make the particle surface unstable, it starts to deform and stretch in one direction leading to the formation of a neck. Due to further oscillations this neck may contract and finally break up into two or more daughter particles, when the fluid particle is exposed to disruptive stresses that exceed the cohesive stresses. Assuming that there is no difference in the mean velocities of the two phases, the disruptive stress is given by the dynamic pressure fluctuations caused by turbulent structures within a given interval. Turbulent structures larger than the fluid particle are commonly assumed to only cause convective movement of the particles and consequently do not contribute to breakage. On the other hand, the turbulent structures much smaller than the fluid particle have not enough energy to break it. Hence, generally to determine the magnitude of the dynamic pressure fluctuations of these turbulent structures, it is assumed that they fall within the inertial subrange. In this case, the disruptive dynamic pressure fluctuations of a turbulent structure of a given size are proportional to: $\rho_c \frac{\bar{u}_\lambda^2}{2}$, where ρ_c is the density of the continuous phase, and $\bar{u}_\lambda^2$ is the mean-squared velocity difference over the distance λ. According to classical turbulence theory, the mean-squared velocity difference in the inertial subrange is given by (Kolmogorov, 1941):

$$\bar{u}_\lambda^2 \approx c_1 (\epsilon\lambda)^{2/3}, \tag{2}$$

where c_1 is a universal constant equal to approximately two, whereas ϵ is the turbulent dissipation rate. Among the stresses counteracting the disruptive pressure term, the interfacial stress and the internal viscous stress are generally the most important. The interfacial stress is proportional to: σ/ξ, where ξ is the equivalent diameter of a sphere with the same volume of the fluid particle, whereas σ is the interfacial tension between the continuous phase and the disperse phase. The internal viscous stress is usually assumed proportional to: $\frac{\mu_d}{\xi}\sqrt{\frac{\tau_{ext}}{\rho_d}}$, where μ_d represents the dynamic viscosity of disperse phase, ρ_d its density and τ_{ext} is the external stress from the continuous phase, which attempts to deform the particle. The internal viscous stress is proportional to the disperse phase viscosity and inversely proportional to the size of the fluid particle. Hence, this can significantly contribute to the stabilization of small viscous liquid drops, while is negligible for bubbles or large drop (Andersson and Andersson, 2006). Considering only the interfacial stress as cohesive stresses in the stress balance, a critical Weber number, above which the fluid particle becomes unstable, can be estimated (Kocamustafaogullari and Ishii, 1995):

$$\mathrm{We}_{cr} = \tau_{ext} \frac{\xi_{max}}{2\sigma} \geq 1. \tag{3}$$

The breakage mechanism described earlier and sketched in Fig. 1 is known as turbulent fluctuation mechanism and is the most common one in industrial equipment. Less important breakage mechanisms include viscous shear stress (B), turbulent shear (C), interfacial instability, and shearing-off, as shown in Fig. 1.

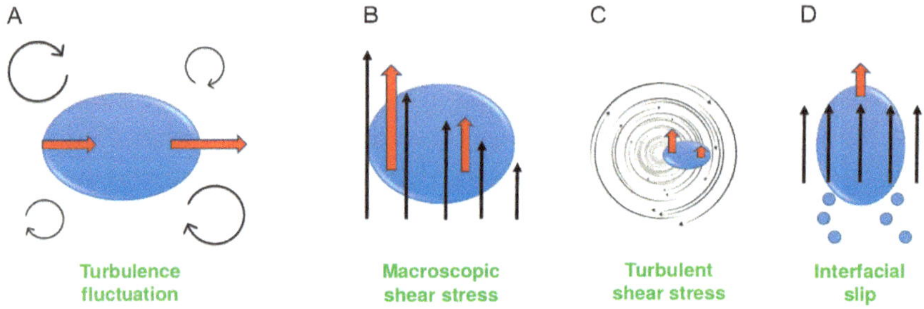

Fig. 1 Different mechanisms leading to fluid particle deformation and breakup in a turbulent flow. (A) Turbulence fluctuation breakup mechanism, (B) macroscopic shear stress breakup mechanism, (C) turbulent shear stress breakup mechanism, (D) interfacial slip breakup mechanism.

Viscous shear forces in the continuous phase will cause a macroscopic velocity gradient around the interface and deform the fluid particle, leading to breakup. If a trailing fluid particle has its larger part outside a wake region, the shear stress across the wake boundary may split it as a result of elongation, fluid particle surface indentation, and necking. The fluid particle is firstly elongated into two lumps separated by a thread and breaks into two almost equally sized daughters corresponding to the lumps and a series of smaller particles called satellites, corresponding to the thread. In the viscous case, the breakage mechanism can be expressed as a balance between external viscous stresses and surface tension forces. The force balance is usually expressed in terms of the dimensionless capillary number:

$$\mathrm{Ca} = \frac{\mu_c \xi \dot{\gamma}_{\max}}{2\sigma}, \tag{4}$$

where ξ is the characteristic size of the fluid particle and $\dot{\gamma}_{\max}$ is the maximum velocity gradient in the external flow field. If the capillary number or fluid particle size exceeds a critical value, the interfacial tension forces are no longer able to maintain the form of the fluid particle, and it will rupture into two or more daughter fluid particles (Kocamustafaogullari and Ishii, 1995; Renardy and Cristini, 2001a,b; Renardy et al., 2002).

If the characteristic dimension of the fluid particle is comparable with the characteristic Kolmogorov length, the breakup can be caused by velocity gradient in the eddies (microscopic velocity gradient (C)). However, even in the absence of net flow of the continuous phase, such as rising bubbles in a stagnant liquid, or falling drops in a continuous gas or immiscible liquid, breakage can be caused by interfacial instabilities due to the Rayleigh–Taylor and Kelvin–Helmholtz instabilities. The Rayleigh–Taylor instability occurs when a light liquid is accelerated into a heavy fluid due to density difference. When the density ratio is approximately of unity, the Kelvin–Helmholtz instability may become important (Kocamustafaogullari and Ishii, 1995).

In a gas–liquid system the shearing-off process is characterized by a number of small fluid particles sheared-off from a large one, also known as erosive breakage. The shearing-off is determined by the balance between the viscous shear force and the surface tension at skirts of the cap/slug fluid particles. When the relative velocity is high enough, the fluid particle skirt is transformed into unsteady sheds, with the generation of a large number of small particles at the rim (Fu and Ishii, 2003a). However, compared to turbulent fluctuations, the influence of viscous shear, shearing-off, and surface instability in a turbulent flow is often negligible and the corresponding theory and modeling is rare in the literature.

The presence of surfactants, salts, or more generally impurities enhances breakage, and as a consequence reduces the particle size. The main reason for this in the case of surfactants is the reduction of interfacial tension, caused by the absorption of molecules at the interface. In addition, when surfactants are present at the interface, pressure fluctuations remove the absorbed molecules thereby exposing a fresh interface, with a dynamic interfacial tension higher than the static one. Thus an extra stress is generated, which adds to the stress due to the turbulent pressure fluctuations. This effect becomes negligible for fluid particles characterized by high density, in which the reduction of the interfacial tension caused by the absorption of surfactants is the major effect (Koshy et al., 1988).

Salts have normally little effect on interfacial tension, but if also surfactants are present in the system, the interfacial tension decreases drastically. The presence of salts enhances the adsorption of surfactant molecules (Kumar et al., 2006), resulting in lower interfacial tension and consequently higher breakage frequencies.

3. PHENOMENOLOGY OF FLUID PARTICLE COALESCENCE

Together with breakage and mass transfer, coalescence controls the evolution of PSD in multiphase flows. Compared to breakage, that involves the interaction between a single fluid particle and the continuous phase, coalescence is intrinsically more complex, as it involves not only interactions of fluid particles with the surrounding, but also between particles themselves, once they are put in contact by external flow or body force (Chesters, 1991).

The first step toward quantitative modeling of coalescence is the choice of a conceptual framework which splits this complex phenomenon into a number of coupled individual subprocesses: (1) collisions and (2) interaction of two or more fluid particles.

The flow field of the continuous phase governs the frequency, intensity, and the duration of the collisions and consequently these aspects influence the following interactions between the particles. Collisions depend on the possible mechanisms with which the fluid particles come into close proximity (see Fig. 2).

Generally, in a turbulent flow collisions between fluid particles are caused by turbulent fluctuations (Fig. 2A) and the eddy capture mechanisms (Fig. 2B). The first is related to the chaotic motion of fluid particles in turbulent flows, caused in turn by the collision with turbulent eddies in the inertial subrange. In fact, the smallest eddies do not contain sufficient energy to affect fluid particles motion, whereas much larger eddies only transport particles and have no effect on their relative motion (Chesters, 1991; Coulaloglou and Tavlarides, 1977; Tsouris and Tavlarides, 1994).

The eddy capture mechanism (Liao and Lucas, 2010) affects fluid particles with size smaller than that falling in the dissipation energy subrange. In this case, collisions are dominated by viscous forces (Chesters, 1991) and in particular the collision frequency depends on the local shear rate of the flow inside turbulent eddies (see Fig. 2B). In addition to the mechanisms described so far, in a turbulent flow other collision including phenomena can be observed, such as collision due to macroscopic velocity gradients,

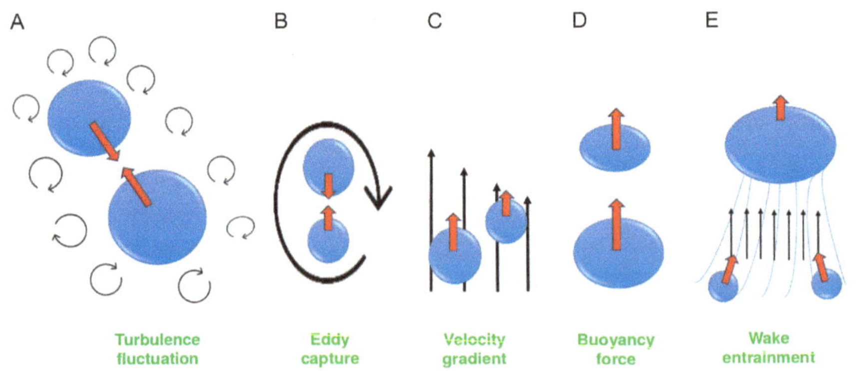

Fig. 2 Possible collision mechanisms for fluid particles in a multiphase turbulent flow. (A) Turbulent fluctuation collision mechanism, (B) eddy capture collision mechanism, (C) velocity gradient collision mechanism, (D) buoyancy force collision mechanism, (E) wake entrainment collision mechanism.

buoyancy force, and wake entrainment. The velocity gradient mechanism (Fig. 2C) indicates collisions induced by relative velocity differences (Chesters, 1991; Liao et al., 2015). This mechanism can be observed also in turbulent flows, where the macroscopic velocity gradients are normally present in some zones of the system. Two fluid particles with different sizes have different rising velocities due to the different contributions of the buoyancy force and can therefore collide (see Fig. 2D) as described by Prince and Blanch (1990). The wake entrainment collision mechanism (Fig. 2E) is important only in gas–liquid systems with large fluid particles. In fact, the motion of these in the continuous phase drags a little amount of liquid, producing a wake behind them. Therefore, the smaller fluid particles close to the wake can be accelerated, caught up and brought to collide with the leading fluid particle, generating the wake (Fu and Ishii, 2003a; Sun et al., 2004).

Not all the collisions lead to coalescence, in fact the fluid particles may separate from each other after the collision. To account for this, an efficiency is introduced. Three different conceptual approaches have been used to derive the coalescence efficiency: energy model, critical approach velocity model, and the film drainage model.

According to the energy model, the fraction of collisions that leads to coalescence depends on the ratio of kinetic collision energy to fluid particle interfacial energy. The kinetic collision energy is proportional to the relative velocity of the two colliding fluid particles and their average volume, whereas the interfacial energy is proportional to the fluid particle surface area and interfacial tension (Sovová, 1981). In the critical approach velocity model, the outcome of a collision depends on the relative velocity of the two fluid particles perpendicular to the surface contact and coalescence occurs only when the approach velocity does not overcome a critical value (Lehr et al., 2002). According to the film drainage model, coalescence between two fluid particles can be divided into three subprocesses (Chesters, 1991; Coulaloglou and Tavlarides, 1977; Narsimhan, 2004; Tsouris and Tavlarides, 1994):

1. two fluid particles collide, trapping a small amount of liquid between them. The film liquid produce a lubrication force that opposes the various attractive and compression forces (i.e., van der Waals electrostatic forces, hydrophobic forces, and fluid compression forces);
2. the attractive forces keep in contact the fluid particles until the liquid drains out down to a critical thickness. At this stage, repulsive forces of interfacial origin, such as electrostatic double layer force, prevent

the fluid particles from coming closer, in addition to the hydrodynamic resistance;

3. in the last stage, when the thickness reaches a critical value (usually of the order of magnitude of nanometers), the van der Waals force causes the film to rupture, and finally coalescence occurs.

In other words, coalescence can occur if the contact time is greater than the drainage time, that is, the time required by the liquid film to reach a critical thickness. Coulaloglou and Tavlarides (1977) obtained the following expression for the coalescence efficiency, assuming that the drainage time is not distributed and the contact time a random variable:

$$\eta = \exp\left(-\frac{\tau_{\text{drainage}}}{\tau_{\text{contact}}}\right). \tag{5}$$

The film drainage model is the most used approach to describe coalescence efficiency, because it allows to describe accurately the interactions phenomena between the fluid particles and the continuous phase after the collision. Moreover, with this model it is possible to consider the effect of contamination. As well as in breakage (see Section 2), both surfactants and salts influence coalescence. Surfactants tend to adsorb at the interface, and during collisions the liquid film is trapped and tends to drain with surfactant molecules swept toward the rim of the film (also known as "barrier ring"). Additionally in the collision, a dimple forms in the film such that the thickness of the film is maximum at the center and minimum at the periphery (see Fig. 3). Therefore, the accumulation of surfactant molecules in the film periphery causes different phenomena which tend to inhibit coalescence. The first concerns the emergence of additional repulsive forces due to the

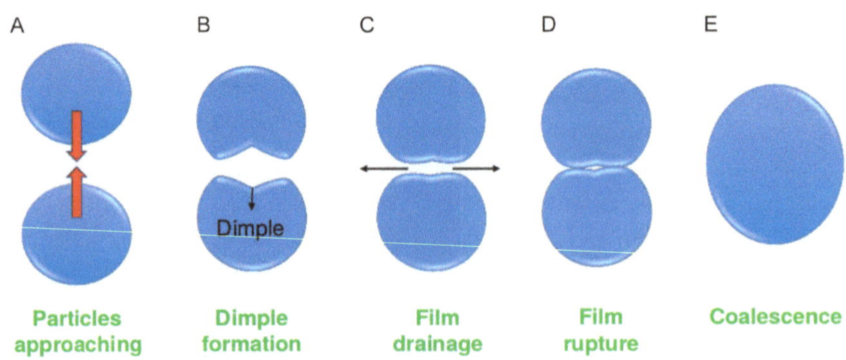

Fig. 3 Sequence of events during the coalescence in according to film drainage model.

steric conformation of surfactants. The repulsion is largest at the barrier ring, where the separation between the two fluid particles (i.e., the thickness of the film) is minimum. Furthermore the surface concentration gradient, caused by a nonuniform distribution of surfactant molecules at the interface, leads to a surface tension gradient, which in turn creates a tangential stress (Gibbs–Marangoni effect) opposing the film drainage and immobilizing the interface. In other words the Gibbs–Marangoni effect increases the drainage time, and so reduces the coalescence frequency. The presence of salt can inhibit or promote the coalescence of fluid particles due to different factors:

- salts reduce the decay length and strength of hydrophobic attractions and thereby retard the drainage of liquid film (Craig et al., 1993);
- as in the case of surfactants, a nonuniform distribution of ions at the interface induces a Gibbs–Marangoni effect, and so the interface becomes less mobile and drainage time increases (Henry et al., 2007; Prince and Blanch, 1990). This happens because the drained film profile tend to have a parabolic shape (Poiseuille flow), and a lower flow rate, than uncontaminated systems, in which the interface is mobile and the velocity profile of the drained film is uniform (plug flow);
- salts reduce the solubility of gas molecules in water solution. Normally, these are present in solution as nano-bubbles with a typical diameter of around 15–20 nm. In a gas–liquid system, nano-bubbles in a thin liquid film migrate toward the interface, leading to an increase of the local disruption of hydrogen bonding and a decrease of the effective viscosity of liquid film. Therefore, in the presence of dissolved gas molecules, the liquid film drains faster due to this increase in interface mobility (Nguyen and Nguyen, 2010).

However, it is also known that some salts have a greater effect on the inhibition of coalescence than others. In particular, the salts in which one of the ions tends to be absorbed in the interface and the other one remains in the bulk (Henry et al., 2007). Furthermore, it has been observed that the salts and ionic-surfactants have a synergistic effect on the inhibition of coalescence (Kumar et al., 2006). In fact, the presence of salts reduces the electrostatic double layer repulsion between the surfactant ions, which increase their adsorption at the interface. On the other hand, in some cases the salts may enhance the coalescence phenomena, because, in according to DLVO theory, the length scale of electric repulsion interactions decreases with the growing of salts concentration.

4. ANALYSIS OF DIFFERENT MODELS FOR FLUID PARTICLE BREAKAGE KERNEL

As discussed in Section 2, numerous models are available to take into account the different breakage mechanisms in the breakage kernel, $b(\xi)$, of fluid particles. In this work only the principal models will be analyzed, and the modification to consider the contamination effect will also be discussed. We will mostly focus on the turbulent fluctuations breakage mechanism, because normally this is the prevailing one in liquid–liquid or gas–liquid industrial systems.

4.1 Turbulent Fluctuations Breakage Models

At least four different criteria can be used to predict breakage according to the turbulent fluctuation mechanism (Liao and Lucas, 2009). A fluid particle breaks:

1. if its *turbulent kinetic energy* becomes greater than a critical value (Coulaloglou and Tavlarides, 1977);
2. if the *velocity fluctuation* around its surface becomes greater than a critical value (Alopaeus et al., 2002; Narsimhan et al., 1979);
3. if the *turbulent kinetic energy* of the bombarding eddies becomes greater than a critical value (Lee et al., 1987a,b; Luo and Svendsen, 1996; Martinez-Bazan et al., 1999a,b; Prince and Blanch, 1990; Tsouris and Tavlarides, 1994);
4. if the *turbulent inertial stress* of hitting eddies becomes greater than the interfacial force of the smallest daughter particle (Lehr and Mewes, 1999; Lehr et al., 2002);

All these breakage models were derived for systems in which turbulence was considered as homogeneous and locally isotropic (Kolmogorov, 1941). Furthermore, all these models describe fluid particle breakage with size falling inside the inertial subrange. Alternatively, Baldyga and Podgórska (1998) derived a fluid particle breakage kernel, applying a multifractal approach to describe the fine-scale structure of turbulence, and accounting for intermittency. Recently, another different approach was developed by Solsvik and Jakobsen (2016) to extend the breakage and coalescence kernels to the complete energy spectrum of isotropic turbulence and to account for low Reynolds number effects.

4.1.1 Turbulent Kinetic Energy for a Fluid Particle Greater Than a Critical Value

One of the first phenomenological models was proposed by Coulaloglou and Tavlarides (1977). Their main assumption was that an oscillating deformed drop can break if the turbulent kinetic energy $E_{\mathrm{d}}(\xi)$ transmitted to fluid particle by turbulent eddies exceeds the fluid particle surface energy E_{cr}. Therefore, the breakage kernel of a fluid particle of size ξ was defined as

$$b(\xi)=\left(\frac{1}{\text{breakage time}}\right)\times\left(\frac{\text{fraction of}}{\text{drops breaking}}\right)=\frac{1}{t_b}\frac{\Delta N(\xi)}{N(\xi)}, \tag{6}$$

where $N(\xi)$ is the total number of particles of size ξ. The fraction of breaking fluid particles was thus expected to be proportional to the fraction of turbulent eddies colliding with the fluid particle that have a turbulent kinetic energy greater than the fluid particle surface energy. Furthermore, it was also assumed that the kinetic energy distribution of the fluid particles is the same as that of eddies with equivalent size. Coulaloglou and Tavlarides (1977) assumed that the motion of fluid particles and eddies was similar to the random motion of gas molecules and consequently they used the two-dimensional space Maxwell–Boltzmann distribution of speed (Kennard, 1938):

$$P(u)\mathrm{d}u=\exp\left(-\frac{u^2}{\bar{u}^2}\right)\mathrm{d}\left(\frac{u^2}{\bar{u}^2}\right), \tag{7}$$

where u is the speed of the simple fluid element and $\bar{u}$ is its averaged value. Considering a fluid particle with surface energy $E_{\mathrm{cr}}=c_1\sigma\xi^2$, the fraction of fluid particles with turbulent kinetic energy larger than E_{cr} equals the number fraction of fluid particles that have velocities greater than a corresponding critical value u_{cr}. This fraction of fluid particles can be estimated through the kinetic energy distribution function as:

$$\frac{\Delta N(\xi)}{N(\xi)}=\int_{E_{\mathrm{cr}}}^{\infty}P(E)\mathrm{d}E=\exp\left(-\frac{E_{cr}}{\bar{E}}\right), \tag{8}$$

where $\bar{E}$ is the mean turbulent kinetic energy:

$$\bar{E}=c_2\rho_d\xi^3\bar{u}^2(\xi) \tag{9}$$

in which c_2 is a dimensionless constant, ρ_{d} is the disperse phase density, and $\bar{u}^2(\xi)$ is the mean-squared relative velocity between two points separated by a distance ξ in the inertial subrange, that according to Kolmogorov (1941) can be written as:

$$\bar{u}^2(\xi) = c_3\epsilon^{2/3}\xi^{2/3}. \tag{10}$$

where c_3 is a universal constant approximately equal to two. The breakage time, t_b, was determined by assuming that the motion of the center of mass of daughter fluid particles is similar to the relative motion of two lumps of fluid in a turbulent flow field. Hence, Coulaloglou and Tavlarides (1977) supposed that the breakage time is given by turbulent eddy turnover time:

$$t_b \propto \xi^{2/3}\epsilon^{-1/3}. \tag{11}$$

Substituting Eq. (9) into Eq. (8), and hence Eqs. (8) and (11) into Eq. (6), the breakage frequency was expressed as:

$$b(\xi) = c_4\xi^{-2/3}\epsilon^{1/3}\exp\left[-\frac{c_5\sigma}{\rho_d\epsilon^{2/3}\xi^{5/3}}\right], \tag{12}$$

where c_4 and c_5 are two empirical parameters that need to be determined experimentally. To take into account the damping effect of fluid particles on local turbulent intensities at high disperse phase hold-up, Coulaloglou and Tavlarides (1977) modified the original expression as:

$$b(\xi) = c_4\xi^{-2/3}\frac{\epsilon^{1/3}}{(1+\alpha_d)}\exp\left[-\frac{c_5\sigma(1+\alpha_d)^2}{\rho_d\epsilon^{2/3}\xi^{5/3}}\right], \tag{13}$$

where α_d is the volume fraction of the disperse phase.

This model derived for liquid–liquid dispersion was also used for gas–liquid mixtures. As pointed out by Prince and Blanch (1990), when applied to gas–liquid systems, the Coulaloglou and Tavlarides (1977) model predicts a breakage kernel several orders of magnitude lower than experimental observations for gas–liquid systems. This might be caused by the fact that the density of the disperse phase in gas–liquid mixture is much lower than that in a liquid–liquid dispersions. For this reason, the density in Eqs. (12) and (13) should be substituted with the continuous phase density (Lasheras et al., 2002).

In general, the Coulaloglou and Tavlarides (1977) breakage model has been criticized because the predicted profile has a maximum value for certain parent drop size (Solsvik et al., 2013). In fact, the forces counteracting fluid particle breakage are inversely proportional to the parent particle size, and thus a realistic breakage model should predict a monotonically rising breakage frequency function with increasing size of the parent. On the other hand, it can be noted that the proposed breakage kernel function may give a maximum (with increasing parent particle size), because this function is inversely proportional to t_b which decreases with the increase of the fluid particle size, while the breakage probability function is increasing (Solsvik et al., 2013).

4.1.2 Turbulent Velocity Fluctuation Around the Particle Greater Than a Critical Value

A stochastic model for the prediction of breakage probability of fluid particles in agitated liquid–liquid dispersions was proposed by Narsimhan et al. (1979). The model was obtained for dispersions with comparable viscosity between the two phases and low volume fraction of the disperse phase. It was assumed that fluid particle breakage is caused by relative velocity fluctuations between two points close to the fluid particle surface. In other words, fluid particle breakage can occur if the turbulent motion of the continuous phase provides the minimum increase in the surface energy needed to break the fluid particle.

As in the case of Coulaloglou and Tavlarides (1977), turbulence was considered homogeneous and locally isotropic. Furthermore, the destructive viscous stresses have been neglected, because it was supposed that in an agitated vessel the Reynolds number is so high that the characteristic turbulence microscale length is much smaller than the size range of drops encountered. Therefore, a fluid particle of size ξ may break due to the inertial forces, transmitted by an eddy of the same or smaller size, which belongs to the inertial subrange. Moreover, it was also assumed that the time-scale of drop oscillation is smaller than the corresponding time-scale associated with the eddy arrival, so that once an eddy of sufficient high energy arrives, the drop would instantly break.

Considering a small time interval, τ, Narsimhan et al. (1979) defined two events: (A) an eddy arrives on the surface of a fluid particle in the time interval τ, and (B) the arriving eddy has energy greater than or equal to the minimum increase in the surface energy, $E_{\min}(V)$ required to break a fluid particle of volume V. The probability that a fluid particle of volume V breaks in the time interval τ equals $P(\mathrm{A}) \times P(\mathrm{B}|\mathrm{A})$, where $P(\mathrm{B}|\mathrm{A})$ denotes the conditional probability of event B, if A is known to occur. Defining λ as an average number of eddies arriving on the fluid particle surface per unit time, the probability with which eddies arrives at the drop surface is

$P(\mathrm{A}) = \lambda\tau$. Therefore, the breakage frequency of a fluid particle of diameter ξ can be defined as follows:

$$b(\xi) = \frac{P(\mathrm{A}) \times P(\mathrm{B}|\mathrm{A})}{\tau} = \lambda P(\mathrm{B}|\mathrm{A}), \quad (14)$$

where λ was considered to be a constant. Narsimhan et al. (1979) hypothesized that binary symmetric breakage entailed the minimum increase in the surface energy. Therefore, being $E_s = \sigma\pi^{1/3}6^{2/3}V^{2/3}$ the surface energy of a fluid particle of volume V, the minimum relative velocity of oscillation required for binary symmetric breakup is

$$E_{\min} = \frac{1}{2}\rho_c V u_{\min}^2 = (2^{1/3} - 1)\sigma\pi^{1/3}6^{2/3}V^{2/3}, \quad (15)$$

hence, a fluid particle breaks if the relative velocity between two points of its surface: $u \geq u_{\min}$.

It was also assumed that the distribution of the relative velocities is reasonably represented by a normal probability density distribution function of the form:

$$P(u) = \frac{1}{\sqrt{2\pi\bar{u}^2}} \exp\left(-\frac{u^2}{2\bar{u}^2}\right), \quad (16)$$

where the relative velocity variance was approximated by $\bar{u}^2(\xi) \approx 2\epsilon^{2/3}\xi^{2/3}$. The conditional probability was thus expressed as:

$$P(\mathrm{B}|\mathrm{A}) = P(u^2 \geq u_{\min}^2) = \int_{u_{\min}}^{\infty} P(u)\mathrm{d}u = \mathrm{erfc}\left[\frac{u_{\min}}{\sqrt{2\bar{u}^2}}\right]. \quad (17)$$

Finally, replacing Eq. (17) into Eq. (14) and using fluid particle diameter ξ as the internal coordinate, the following expression is obtained

$$b(\xi) = \frac{\lambda}{2}\mathrm{erfc}\left[\sqrt{2(2^{1/3} - 1)}\sqrt{\frac{\sigma}{\rho_c}\frac{6^{-1/18}\xi^{-5/6}}{2\epsilon^{1/3}}}\right]. \quad (18)$$

It can be noted that Eq. (18) does not depend on empirical constants. On the other hand, it is worth pointing out that in this model, the surface energy increase was assumed to be at its minimum when the parent fluid particle breaks into two equally sized daughter fluid particles. This assumption is, however, inconsistent with experimental evidence. In fact, Hesketh et al. (1987) showed that binary equal-sized symmetric breakage has the lowest probability to occur. This observation was confirmed by Andersson and

Andersson (2006), which proved as for drops, multiple breakage is the most common event, whereas the binary equal-sized breakage occurs for bubbles under specific conditions.

Alopaeus et al. (2002) modified the Narsimhan et al. (1979) model, introducing a particle-eddy collision frequency parameter, λ, that depends on turbulent dissipation rate, ϵ, with a $1/3$ power. Furthermore, Alopaeus et al. (2002) also included the viscous energy term in the kinetic energy balance (Eq. 15), to take into account the resistance opposed by internal viscous stresses. The model elaborated has the following expression:

$$b(\xi) = c_6 \epsilon^{1/3} \text{erfc}\left[\sqrt{c_7 \frac{\sigma}{\rho_c \epsilon^{2/3} \xi^{5/3}} + c_8 \frac{\mu_d}{\sqrt{\rho_c \rho_d} \epsilon^{1/3} \xi^{4/3}}}\right], \tag{19}$$

where c_6 is a fitting parameter having specific dimensions, whereas c_7 and c_8 are dimensionless fitting parameters.

Laakkonen et al. (2006) proposed a variant of this breakage frequency to adapt it also to gas–liquid systems. They substituted the disperse phase viscosity with the continuous one, following the experimental observation of Walter and Blanch (1986). It is worth mentioning that c_6 is not dimensionless and therefore some scale dependence might be included, suggesting that c_6 is not a universal constant.

4.1.3 Turbulent Kinetic of the Bombarding Eddy Greater Than a Critical Surface Energy Value

As in the work of Narsimhan et al. (1979), Lee et al. (1987a,b) proposed a stochastic model for binary breakage, assuming local isotropic turbulence for high Reynolds numbers and neglecting viscous forces. Differently from Narsimhan et al. (1979), Lee et al. (1987a,b) supposed that particle breakage occurs if the kinetic energy of the arriving eddies overcomes a defined value proportional to the particle energy surface. They, considering a small time interval τ, defined five events: (A) one eddy arrives on the fluid particle surface in the time interval τ, (B) the arriving eddy has kinetic energy greater than or equal to the minimum increase in surface energy required for the breakage of the fluid particle of size ξ, (A') more than one eddy arrive on the fluid particle surface in the time interval τ, (B') at least one arriving eddy has a kinetic energy greater than or equal to the minimum required for fluid particle breakage, and (D) a fluid particle with size ξ breaks in the time interval τ. Hence, they defined the breakage kernel of a fluid particle of diameter ξ as

$$b(\xi) = P(\mathrm{D}) = P(\mathrm{A})P(\mathrm{B}|\mathrm{A}) + P(\mathrm{A}')P(\mathrm{B}'|\mathrm{A}') \approx P(\mathrm{A})P(\mathrm{B}|\mathrm{A}) = \lambda P(\mathrm{B}|\mathrm{A}), \tag{20}$$

in fact if τ is small enough, event A′ becomes much less likely than A.

Since only an eddy that has length-scale comparable to particle size ξ can cause breakage, it was supposed that the conditional probability density of the arriving eddy with length-scale l_e, given that an eddy arrives on the surface, $f(l_e|\mathrm{A})$, is uniformly distributed over the interval $(0, \xi)$: $f(l_e|\mathrm{A}) = 1/\xi$ for $0 \leq l_e \leq \xi$. Therefore, the breakage kernel is expressed as

$$b(\xi) = \lambda \int_0^{\xi} P(\mathrm{B}|l_e) f(l_e|\mathrm{A}) \mathrm{d}l_e = \frac{\lambda}{\xi} \int_0^{\xi} P(\mathrm{B}|l_e) \mathrm{d}l_e, \tag{21}$$

where $P(B|l_e)$ is the conditional probability that an eddy of length-scale l_e arrives on the fluid particle surface with a kinetic energy greater than or equal to the minimum energy required for fluid particle breakage. In this case, unlike Narsimhan et al. (1979), the collision frequency, λ, was assumed dependent on the turbulent dissipation rate ϵ and the size ξ through a power law:

$$\lambda = C_\lambda \epsilon^{1/2} \xi^{2/3}, \tag{22}$$

where C_λ is an empirical parameter, whereas the exponents of ϵ and ξ were obtained by dimensional analysis. The conditional probability $P(\mathrm{B}|l_e)$ was evaluated assuming that the kinetic energy distribution of turbulent eddies follows the Maxwell–Boltzmann distribution (Kennard, 1938):

$$K(x) = \frac{1}{\sqrt{2\pi}} x^{1/2} \exp\left(\frac{x}{2}\right), \tag{23}$$

where $x = 3E(l_e)/\bar{E}(l_e)$ and, $E(l_e)$ denotes the kinetic energy of the particular eddy of length-scale l_e, and $\bar{E}(l_e)$ is the mean kinetic energy of the turbulent eddies which have length-scale l_e. The condition for breakage used by Lee et al. (1987a,b) is that breakage occurs if the kinetic energy of the arriving eddy exceeds a certain fraction of the particle surface energy: $E(l_e) \geq c_9 \pi \xi^2 \sigma$. According to the breakage condition, the conditional probability can be expressed as:

$$P(\mathrm{B}|l_e) = 1 - P(x \leq X_c) = 1 - F(X_c), \tag{24}$$

where $F(\cdot)$ is the cumulative chi-square distribution with three degrees of freedom, and $X_c = 3c_9\pi\xi^2\sigma/\bar{E}(l_e)$ is the critical value of x. Replacing Eqs. (22) and (24) into Eq. (21), the final expression is obtained

$$b(\xi) = C_\lambda \epsilon^{1/3}\xi^{-2/3}\left[1 - \frac{1}{\xi}\int_0^\xi F\left(\frac{c_{10}\sigma\xi^2}{\rho_c\epsilon^{2/3}l_e^{11/3}}\right)\mathrm{d}l_e\right], \tag{25}$$

where the two empirical constant C_λ and c_{10} must be determined experimentally.

Compared with Narsimhan et al. (1979) model, Lee et al. (1987a,b) have added the reasonable ϵ and ξ dependencies of collision frequency, λ, and removed the inconsistent hypothesis on the equal binary breakage. On the other hand, this model is very sensitive to the value of the parameter c_{10} (Liao and Lucas, 2009) that represents the minimum ratio of the kinetic eddy energy to the surface energy required for breakage. However, this model is less used than Coulaloglou and Tavlarides (1977) and Narsimhan et al. (1979) models due to required solution of the integral in Eq. (25).

Another phenomenological model, based on the direct interaction between the particles and the turbulent eddies, was proposed by Prince and Blanch (1990) who considered the breakage frequency as product of collision frequency $w(\xi, l_e)$, and collision efficiency $P_b(\xi, l_e)$ resulting in:

$$b(\xi) = \int_0^\infty w(\xi, l_e)P_b(\xi, l_e)\mathrm{d}l_e. \tag{26}$$

According to gas kinetic theory, the collision frequency was defined as the volume swept in unity time, which is equal to the product of the particle-eddy cross sectional area $S(\xi, l_e)$,the particle-eddy relative velocity $u_{rel}(\xi, l_e)$ and the number density of eddies with l_e, n_e:

$$w(\xi, l_e) = n_e S(\xi, l_e)u_{rel}(\xi, l_e). \tag{27}$$

Prince and Blanch (1990) assumed turbulence isotropic, eddy size falling in the inertial subrange resulting in a number of eddies equal to:

$$\frac{\mathrm{d}N(\kappa)}{\mathrm{d}\kappa} = \frac{0.1\kappa^2}{\rho_c}, \tag{28}$$

where $N(\kappa)$ is the number of eddies of wave number $\kappa = 2/l_e$ per mass of fluid. Prince and Blanch (1990) pointed out that the number of eddies predicted by Eq. (28) becomes infinitely large as the eddy size approaches zero. Therefore, they considered an arbitrary size (i.e., smaller 20% than particle size), below which an eddy cannot cause breakage. However, it was noted as the simulation results are sensitive to this limit value (Lasheras et al., 2002). The collision efficiency $P_b(\xi, l_e)$ was defined similarly to Coulaloglou and Tavlarides (1977):

$$P_b(\xi, l_e) = \exp\left(-\frac{E_{cr}}{\bar{E}}\right) \approx \exp\left(-\frac{We_{cr}\frac{\sigma}{\rho_c}\xi}{2(\xi l_e)^{2/3}}\right). \tag{29}$$

As it can be seen from Eq. (29), the critical energy required for breakage to occur is determined from the critical Weber number, which for a bubble is usually taken equal to $We_{cr} = 2.3$. Finally, using Eq. (27) and Eq. (29), the breakage kernel is then given by:

$$\begin{aligned} b(\xi) &= \int_{\kappa_{min}}^{\kappa_{max}} \frac{0.14\pi}{16}\left(\xi + \frac{2}{\kappa}\right)^2 \left(\xi^{2/3} + \left(\frac{2}{\kappa}\right)^{2/3}\right)^{1/2} \\ &\quad \times \epsilon^{1/3} \exp\left[-\frac{1.18}{(2)^{2/3}} \frac{\sigma\kappa^{2/3}}{\rho_c \xi \epsilon^{2/3}}\right] \kappa^2 d\kappa, \end{aligned} \tag{30}$$

where $\kappa = 2/l_e$ is the wave number of eddies.

Tsouris and Tavlarides (1994) proposed a modification of Prince and Blanch (1990) model for liquid–liquid dispersions employing a different expression for E_{cr} in Eq. (29):

$$E_{cr} = \frac{\pi\sigma}{2}\left[2\left(\frac{\xi}{2^{1/3}}\right)^2 + \xi_{max}^2 + \xi_{min}^2 - 2\xi^2\right], \tag{31}$$

in which ξ_{min} is the minimum fluid particle size and $\xi_{max} = (\xi^3 - \xi_{min}^3)^{1/3}$ is fluid particle size with complementary volume. Hence, the expression of breakage kernel becomes

$$\begin{aligned} b(\xi) &= 0.0118 DF(\alpha_d)\epsilon^{1/3} \\ &\times \int_{\kappa_{min}}^{\kappa_{max}} \left(\frac{2}{\kappa} + \xi\right)^2 (8.2\kappa^{-2/3} + 1.07\xi^{2/3})^{1/2} \kappa^2 \exp\left[-\frac{E_{cr}}{c_{11} 0.43 \rho_c \pi \xi^{11/3} \epsilon^{2/3}}\right] d\kappa, \end{aligned} \tag{32}$$

where $DF(\alpha_d)$ is a turbulent damping factor and c_{11} is an empirical parameter. The turbulent damping factor was introduced by Tsouris and Tavlarides (1994) in the expression of the eddy-drop collision frequency (Eq. 27) to take into account the turbulence damping effect caused by the presence of the disperse phase:

$$DF(\alpha_\mathrm{d}) = 1 + 2.5\alpha_\mathrm{d}\left(\frac{\mu_\mathrm{d} + 0.4\mu_\mathrm{c}}{\mu_\mathrm{d} + \mu_\mathrm{c}}\right)^2. \tag{33}$$

Tsouris and Tavlarides (1994) criticized the Coulaloglou and Tavlarides (1977) model, because it predicts a maximum value with parent size increases, in contrast with same experimental evidence. However, the models of Lee et al. (1987a,b), Prince and Blanch (1990), and Martinez-Bazan et al. (1999a), among others, exhibit the same behavior of Coulaloglou and Tavlarides (1977) breakage frequency that was validated by experimental data some years later (Gäbler et al., 2006; Maaß et al., 2010).

Another popular model, based on direct interaction between the eddies and fluid particles (drops and bubbles), was developed by Luo and Svendsen (1996). Differently from previous models (Prince and Blanch, 1990; Tsouris and Tavlarides, 1994), Luo and Svendsen (1996) proposed a breakage kernel with no empirical constants. As in Prince and Blanch (1990), they proposed a frequency of collisions between turbulent eddies and particles:

$$b(\xi, \xi_\mathrm{d}) = \int_0^\infty w(\xi, l_\mathrm{e}) P_b(\xi, \xi_\mathrm{d}, l_\mathrm{e}) \mathrm{d}l_\mathrm{e}, \tag{34}$$

where $w(\xi, l_\mathrm{e})$ was introduced in Eq. (27), whereas $P_b(\xi, \xi_\mathrm{d}, l_\mathrm{e})$ is the probability for a particle of size ξ to break into two daughter fluid particles, one with size ξ_d, when the particle is hit by an arriving eddy of size l_e. $P_b(\xi, \xi_\mathrm{d}, l_\mathrm{e})$ is the same as in Eq. (29), in which E_cr has been evaluated considering the energy required for a parent particle of diameter ξ to be broken into two daughter particles of size ξ_d and $(\xi^3 - \xi_\mathrm{d}^3)^{1/3}$:

$$E_\mathrm{cr}(\xi, \xi_\mathrm{d}) = \pi\sigma\left(\xi_\mathrm{d}^2 + (\xi^3 - \xi_\mathrm{d}^3)^{2/3} - \xi^2\right) = c_f(\xi, \xi_\mathrm{d})\pi\sigma\xi^2, \tag{35}$$

in which the surface area increase coefficient, $c_f(\xi, \xi_\mathrm{d})$, varies with the daughter particle size, but within the interval from 0 to 0.26. Therefore, the breakage kernel of a particle with size ξ that breaks into two daughters with size ξ_d and $(\xi^3 - \xi_\mathrm{d}^3)^{1/3}$ is given as:

$$\begin{aligned} b(\xi, \xi_\mathrm{d}) &= 0.923(1 - \alpha_\mathrm{d})\left(\frac{\epsilon}{\xi^2}\right)^{1/3} \\ &\times \int_{\zeta_\mathrm{min}}^1 \frac{(1+\zeta)^2}{\zeta^{11/3}} \exp\left[-\frac{12 c_f \sigma}{2\rho_c \epsilon^{2/3} \xi^{5/3} \zeta^{11/3}}\right] \mathrm{d}\zeta, \end{aligned} \tag{36}$$

where $\zeta_{\min} = l_{\mathrm{e,min}}/\xi$ and $l_{\mathrm{e,min}} = (11.4 \sim 31.4)(\mu_{\mathrm{c}}/\rho_{\mathrm{c}})^{3/4}\epsilon^{-1/4}$. It can be noted that Eq. (36) represents a partial breakage kernel, because it takes into account only the breakage of fluid particle with size ξ into two particles of defined size ξ_{d} and $(\xi^3 - \xi_{\mathrm{d}}^3)^{1/3}$. Hence, to consider all possible breakage phenomena of particle of size ξ, it is necessary to integrate Eq. (36):

$$b(\xi) = \frac{1}{2}\int_0^{\infty} b(\xi, \xi_{\mathrm{d}})\mathrm{d}\xi_{\mathrm{d}}, \tag{37}$$

where the factor 1/2 is used not to take into account the same event twice. As aforementioned, this model does not contain unknown or adjustable parameters. On the other hand, Lasheras et al. (2002) point out that this model, as the previous ones (Prince and Blanch, 1990; Tsouris and Tavlarides, 1994), is very sensitive to the specified lower limit of integration. Another advantage of the Luo and Svendsen (1996) model is that no predefined daughter size distribution, must be specified.

The models described so far (Lee et al., 1987a,b; Luo and Svendsen, 1996; Prince and Blanch, 1990; Tsouris and Tavlarides, 1994) were based on the extension of classical kinetic theory of gas as to describe the interaction between the fluid particles and turbulent eddies. Martinez-Bazan et al. (1999a,b) criticized the earlier models in which turbulence was assumed to be composed from discrete eddies, and proposed an alternative model based on purely kinematic ideas. The model was based on a simple balance between the deforming turbulent stress (τ_t) and the surface restoring stresses (τ_s). Moreover, the internal viscous restoring stress was negligible compared to the restoring surface tension stress, and the external deformation viscous stresses compared to the deforming turbulent stress. The restoring stress was defined as the minimum energy required to deform a spherical particle of size ξ per unit bubble volume:

$$\tau_s = 6\frac{\sigma}{\xi}. \tag{38}$$

If the particles are within the inertial subrange of turbulence, the average deformation stress can be estimated as (Kolmogorov, 1941):

$$\tau_t = \frac{1}{2}\rho_{\mathrm{c}}\bar{u}^2, \tag{39}$$

where $\bar{u}$ is the velocity fluctuation in the inertial subrange (Eq. 10). Martinez-Bazan et al. (1999a,b) defined the breakage time as

$$t_b \propto \frac{\xi}{\sqrt{\frac{2\tau_t}{\rho_c} - \frac{2\tau_s}{\rho_c}}}, \tag{40}$$

and consequently, the breakage kernel is defined as the inverse of the breakage time:

$$b(\xi) = c_{12} \frac{\sqrt{2(\epsilon\xi)^{2/3} - 12\frac{\sigma}{\rho_c \xi}}}{\xi}, \tag{41}$$

on which c_{12} must be estimated experimentally.

The Martinez-Bazan et al. (1999a,b) model, even if it is based on a stress balance, can be considered an alternative model in which a fluid particle breaks if the turbulent kinetic energy in the continuous phase is larger than a critical value (Lee et al., 1987a,b; Luo and Svendsen, 1996; Prince and Blanch, 1990; Tsouris and Tavlarides, 1994) because the stress and the kinetic energy balance are not independent. As it can be seen, in this model the probabilistic approach for the distribution density of kinetic energy was not used, representing the mean limitation, because breakage is a probabilistic and a not mechanistic event (Solsvik et al., 2013). Furthermore, this model is not suitable to describe the breakage phenomena in liquid–liquid dispersions with high disperse phase density, because the internal viscous restoring stress is neglected.

4.1.4 Turbulent Inertial Stress of a Hitting Eddy Greater Than the Interfacial Force of the Smallest Daughter Particle

Differently from all models described in the previously paragraphs (excluding the Martinez-Bazan et al. (1999a,b) model), in which only energy constraints were considered, Lehr and Mewes (1999) proposed a model based on the force balance between the inertial force of the arriving eddy and the interfacial restoring force of the smaller daughter particle, also known as *capillary constraint* (Wang et al., 2003). Lehr and Mewes (1999) claimed that the capillary pressure is the dominant constraint for the breakage of fluid particles with radius tending to zero, because in this case the interfacial force is very high and thus the arriving eddy may not provide enough dynamic pressure to overcome the capillary pressure, even though the hitting eddies have enough kinetic energy to fulfill the criteria discussed in Section 4.1.3. This breakage frequency model was originally developed for gas–liquid dispersions.

Lehr and Mewes (1999) used Luo and Svendsen (1996) approach to derive a partial breakage kernel of the smallest daughter; thus Eq. (36) was the starting point of the model development. However, Lehr and Mewes (1999), differently from Luo and Svendsen (1996), assumed that the breakage probability depends on the angle ϕ under which the hitting eddy approaches the particle. They calculated the relative probability from the following average force balance:

$$\frac{\rho_c \bar{u}^2(l_e)}{2}\cos(\phi) = \frac{\sigma}{\xi_{d,min}}, \tag{42}$$

in which $\bar{u}(l_e)$ denotes the velocity of an eddy of size l_e (Eq. 10), while $\xi_{d,min}$, in this case, represents the diameter of the smaller of two daughter particles. On the basis of an average form of the above force balance the collision probability (or collision efficiency) can be obtained, and using an expression similar to Eq. (27) for the collision frequency, $w(\xi, l_e)$, the partial breakage kernel can be expressed as:

$$\begin{aligned} b(\xi,\xi_d) &= \frac{3.65\xi(1-\alpha_d)\sigma}{\rho_c \epsilon^{1/3} \xi_{d,min}^4} \int_{l_{e,min}}^{\xi} \frac{1}{l_e^{10/3}} dl_e \\ &= \frac{1.56\xi(1-\alpha_d)\sigma}{\rho_c \epsilon^{1/3} \xi_{d,min}^4} \left(\frac{1}{l_{e,min}} - \frac{1}{\xi^{7/3}} \right). \end{aligned} \tag{43}$$

Also, this model assumes that only the eddies with size l_e less than or equal to the parent particle can induce breakage; for this reason the upper limit of the integration was fixed to $l_{e,max} = \xi$ in Eq. (43). Unlike previous models (Luo and Svendsen, 1996; Prince and Blanch, 1990; Tsouris and Tavlarides, 1994) the lower limit of integration is not fixed arbitrarily, but estimated as:

$$l_{e,min} = \max\left[\xi_{d,min}, \frac{\sigma^{3/2}}{\rho_c^{3/2} \epsilon \xi_{d,min}^{3/2}}\right], \tag{44}$$

where the second term in the parenthesis is $\xi_{d,min}$ evaluated by the force balance at Eq. (42). As in the case of the Luo and Svendsen (1996) model, Eq. (43) must be integrated for all possible daughter sizes to obtain the breakage kernel. However, this model has the same advantage of Luo and Svendsen (1996) model, because the daughter particle size distribution can be obtained directly from the breakage frequency. On other hand, the Lehr and Mewes (1999) model was criticized because the breakage efficiency was not derived from a cumulative exponential distribution function and it did not satisfy the normalization condition. For this reasons, Lehr et al. (2002) modified the model introducing an exponential distribution function for the breakage kernel, and a critical energy obtained from the following stress balance:

$$\frac{1}{2}\rho_c \bar{u}^2(l_e) = \frac{2\sigma}{\xi_{d,min}}, \tag{45}$$

where $\xi_{d,min}$ is again the diameter of the smaller of the two daughter fluid particles, and $\bar{u}^2(l_e)$ is the velocity of an eddy of size l_e, where for the inertial subrange: $\bar{u}^2(l_e) \approx 2(\epsilon l_e)^{2/3}$. It can be noted from Eq. (45) that Lehr et al. (2002) did not take into account the angle under which the eddy hits the particle, because there is not much information available about the distribution of the eddy-bubble approach angles. In conclusion, starting from Luo and Svendsen (1996) modeling framework and using the modification above described, Lehr et al. (2002) obtained the following expression for the partial breakage kernel:

$$b(\xi, \xi_d) = 1.190 \int_{\xi_{dau,min}}^{\xi} \frac{\sigma}{\rho_c \epsilon^{1/3} \xi_{dau}^4} \frac{(l_e + \xi)^2}{l_e^{11/3}} \exp\left(-\frac{2\sigma}{\rho_c \epsilon^{2/3} l_e^{2/3} \xi_{dau}}\right) dl_e. \tag{46}$$

Instead of integrating Eq. (46) directly, Lehr et al. (2002) expressed the integral as a sum of incomplete gamma-functions to provide an analytical solution for the breakage frequency resulting in:

$$b(\xi) = 0.5 \frac{\xi^{5/3} \epsilon^{19/15} \rho_c^{7/5}}{\sigma^{7/5}} \exp\left(-\frac{\sqrt{2}\sigma^{9/5}}{\xi^3 \rho_c^{9/5} \epsilon^{6/5}}\right). \tag{47}$$

During the last two decades many models have been proposed as extension or combination of the different criteria describe in this section. For example, Wang et al. (2003) and Zhao and Ge (2007) combined the model of Luo and Svendsen (1996) by adding the capillary constraint used by Lehr and Mewes (1999). Andersson and Andersson (2006) modified the original Luo and Svendsen (1996) model to take into account some experimental observation on the high deformation of fluid particle before breakage occurs. Liao et al. (2011) proposed a generalization of the original fluid particle breakage kernel model proposed by Martinez-Bazan et al. (1999a,b), considering the effect of turbulent, viscous and interfacial friction forces. The Martinez-Bazan et al. (1999a,b) breakage model was also adapted by Eastwood et al. (2000) to describe fluid particles breakage in liquid–liquid jet system analyses. The Martinez-Bazan et al. (1999a,b) model has also been combined with the Coulaloglou and Tavlarides (1977) model by Solsvik et al. (2017) to extend the original model for finite Reynolds number. This approach will be examined in greater detail in Section 4.1.6.

4.1.5 Breakage Kernel Multifractal Model

To consider the intermittent effect of turbulence, Baldyga and Podgórska (1998) applied the multifractal theory to derive a modified breakage function for particles with sizes within the inertial subrange.

The Kolmogorov (1941) theory is not complete as it neglects, for example, intermittency, that is, the dynamics of vortex stretching resulting in the formation of regions of space characterized by high rates of energy dissipation, surrounded by nearly irrotational fluid. For this reason, even when the mean value of the rate of energy dissipation $\langle \epsilon \rangle$ results uniformly distributed, there is a large variability in time and in space of its instantaneous and local values. Moreover, the higher-order structure functions for the velocity and the turbulent stresses do not scale as predicted by Kolmogorov (1941). In fact, applying the multifractal theory to describe this phenomenon occurring within the inertial subrange of turbulence, the velocity increment u_r over a distance r scales as:

$$\frac{u_r}{u_L} = \left(\frac{r}{L}\right)^{\frac{\alpha}{3}}, \tag{48}$$

where α is the scaling exponent, L represent the integral scale of turbulence, and u_L is the velocity difference over the distance L. Considering that L and u_L are related through the mean rate of dissipation of kinetic energy as $u_L = [\langle \epsilon \rangle L]^{1/3}$, Eq. (48) can be rewritten as:

$$u_r = u_L\left(\frac{r}{L}\right)^{\frac{\alpha}{3}} = [\langle \epsilon \rangle r]^{\frac{1}{3}}\left(\frac{r}{L}\right)^{\frac{\alpha-1}{3}}. \tag{49}$$

The instantaneous local normal turbulent stresses in the inertial subrange acting on a fluid particle of size ξ can be computed through Eq. (49):

$$\tau_t = \frac{1}{2}\rho_c u_r^2 = \frac{1}{2}\rho_c[\langle \epsilon \rangle \xi]^{\frac{2}{3}}\left(\frac{\xi}{L}\right)^{\frac{2(\alpha-1)}{3}}. \tag{50}$$

In multifractal theory the characteristic frequency of eddies of scale ξ is different for different eddies. Hence, the breakage frequency for a particle of size ξ will depend also the scaling exponent α (Baldyga and Podgórska, 1998):

$$b(\xi,\alpha)=\frac{u_r}{\xi}=\frac{\langle\epsilon\rangle^{\frac{1}{3}}\left(\frac{\xi}{L}\right)^{\frac{\alpha-1}{3}}}{\xi^{\frac{2}{3}}}. \tag{51}$$

Turbulent events, labeled by α, occur with difference probability and a range of possible scaling exponents α, in which each α value corresponds a specific fractal dimension, can be found. The set with scaling exponents between α and $\alpha + d\alpha$ has a fractal dimension $f_\xi(\alpha)$, and therefore the probability density function for α in a box of length r in a ξ_s-dimensional space is given by (Baldyga and Bourne, 1999):

$$P(\alpha)\approx\sqrt{\ln\left(\frac{L}{r}\right)}\left(\frac{r}{L}\right)^{\xi_s-f_\xi(\alpha)}. \tag{52}$$

For high Reynolds numbers, the turbulent flow is assumed to have scaling exponents in the range: $\alpha \in [\alpha_{\min}, \alpha_{\max}]$, where $\alpha_{\min}$, $\alpha_{\max}$, and $f_\xi(\alpha)$ are assumed to be universal (i.e., independent of the large-scale flow). To compute the given probability distribution, the multifractal distribution $f_\xi(\alpha)$ must be specified and generally a parabolic approximation is employed

$$f_\xi(\alpha)=1-\frac{(\alpha-1.117)^2}{0.468}, \tag{53}$$

or in alternative, a polynomial function $f_\xi(\alpha)$, fitted with experimental data (Baldyga et al., 1997) is also employed:

$$f_\xi(\alpha)=a+b\alpha+c\alpha^2+d\alpha^3+e\alpha^4+f\alpha^5+g\alpha^6+h\alpha^7+i\alpha^8, \tag{54}$$

where $a = -3.510$, $b = 18.721$, $c = -55.918$, $d = 120.900$, $e = -162.54$, $f = 131.51$, $g = -62.572$, $h = 16.100$, and $i = -1.7264$, for $\alpha \geq 0.12$.

Considering fluid particles with low viscosity, these are assumed be unstable when the disruptive turbulent stresses become greater than interfacial cohesive stresses:

$$\frac{4\sigma}{\xi}<c_1\rho_c[\langle\epsilon\rangle\xi]^{2/3}\left(\frac{\xi}{L}\right)^{\frac{2(\alpha-1)}{3}}. \tag{55}$$

Nevertheless, for a particle of size ξ, only a fraction of the eddies of scale ξ is vigorous enough to cause particle breakage. These vigorous eddies are labeled by the values of the scaling exponent $\alpha \in [\alpha_{\min}, \alpha_x]$ where:

$$\alpha_x(\xi) = \frac{\frac{5}{2}\ln\left[\frac{L\langle\epsilon\rangle^{0.4}\rho_c^{0.6}}{c_x\sigma^{0.6}}\right]}{\ln(L/\xi)} - 1.5. \tag{56}$$

α_x was obtained from the inequality stress relation (Eq. 55), whereas the lower bound, $\alpha_{\min}$, is the minimum of the scaling exponent range.

The characteristic frequency of eddies of size ξ having the scale exponent α is given by Eq. (51). Such eddies appear with the probability $P(\alpha)\mathrm{d}\alpha$, where $P(\alpha)$ is the probability density function given by Eq. (52). Summing up the contributions to the breakage frequency from all vigorous eddies, the overall breakage frequency can be expressed as:

$$b(\xi) = \int_{\alpha_{\min}}^{\alpha_x} b(\xi,\alpha)P(\alpha)\mathrm{d}\alpha. \tag{57}$$

Using Eq. (52) to express $P(\alpha)$, the final expression for the breakage frequency can be obtained

$$b(\xi) = c_b\sqrt{\ln(L/\xi)}\langle\epsilon\rangle^{1/3}\xi^{-2/3}\int_{\alpha_{\min}}^{\alpha_x}\left(\frac{\xi}{L}\right)^{\frac{\alpha+2-3f_\xi}{3}}\mathrm{d}\alpha. \tag{58}$$

It be can noted from Eq. (58) that there is a dependence on the integral scale of turbulence L, which varies with the investigated system scale. Another feature of Eq. (58) is a very low breakage rate for small particles due to the decrease of α_x with the decrease of ξ (see Eq. 56).

For fluid particles with high viscosity, the viscous restoring stress must be considered. Hence, Eq. (55) must be rewritten as follows:

$$\frac{4\sigma}{\xi} + \mu_\mathrm{d}\frac{\partial u_x}{\partial x} < c_{13}\rho_c[\langle\epsilon\rangle\xi]^{2/3}\left(\frac{\xi}{L}\right)^{\frac{2(\alpha-1)}{3}}, \tag{59}$$

consequently a new expression for α_x (Baldyga and Podgórska, 1998) can be obtained

$$\alpha_x(\xi) = \frac{3\ln\left\{2\left[\frac{\beta_\mu c_x^{5/3}\mu_\mathrm{d}}{\rho_c\langle\epsilon\rangle^{1/3}L^{1/3}\xi} + \sqrt{\left(\frac{\beta_\mu c_x^{5/3}\mu_\mathrm{d}}{\rho_c\langle\epsilon\rangle^{1/3}L^{1/3}\xi}\right) + \frac{4c_x^{5/3}\sigma}{\rho_c\langle\epsilon\rangle^{2/3}\xi}}\right]^{-1}\right\}}{\ln(L/\xi)}, \tag{60}$$

in which $c_x = 0.435$ and $\beta_\mu = 1.91$ are two universal constants. It can be noted that Eq. (60) is a generalization of Eq. (56), in fact when μ_d tends to zero the two equations become identical.

If the fluid particles, in a turbulent dispersed system, fall in the viscous subrange of turbulence in accordance to the Kolmogorov (1941) theory, $\xi < l_{e,\ K}$, the breakage mechanism can be consider viscous. In this case the local value of the turbulent dissipation rate is constant and equal to (Baldyga and Podgórska, 1998):

$$\frac{\epsilon}{\langle\epsilon\rangle} = \left(\frac{\langle l_{e,K}\rangle}{L}\right)^{\frac{4(\alpha-1)}{\alpha+3}}. \tag{61}$$

Using this relationship one gets for the viscous stress:

$$\tau_s = \mu_c\left(\frac{\epsilon}{\nu_c}\right)^{1/2} = \overline{\tau}_s\left(\frac{\langle l_{e,K}\rangle}{L}\right)^{\frac{2(\alpha-1)}{\alpha+3}} \tag{62}$$

where $\overline{\tau}_s = \mu_c(\langle\epsilon\rangle/\nu_c)^{1/2}$. Therefore, the local breakage frequency is taken equal to the local strain S:

$$b(\alpha,\xi)_{\text{visc}} = S = \left(\frac{\epsilon}{\nu_c}\right)^{1/2} = \overline{S}\left(\frac{\langle l_{e,K}\rangle}{L}\right)^{\frac{2(\alpha-1)}{\alpha+3}}, \tag{63}$$

in which $\overline{S} = (\langle\epsilon\rangle/\nu_c)^{1/2}$. In this case the probability density function, function of α, can be approximated as

$$P(\alpha) = c_{14}\left(\frac{\langle l_{e,K}\rangle}{L}\right)^{\frac{4[\xi_s - f_\xi(\alpha)]}{\alpha+3}}, \tag{64}$$

where c_{14} is an empirical parameter. Consequently, accounting for all contributions to the breakage frequency from all vigorous eddies, results in:

$$\begin{aligned} b(\xi)_{\text{visc}} &= \int_{\alpha_{\min}}^{\alpha_x} b(\xi,\alpha)_{\text{visc}} P(\alpha)\mathrm{d}\alpha \\ &= c_{14}\left(\frac{\langle l_{e,K}\rangle}{L}\right)^{1/2} \int_{\alpha_{\min}}^{\alpha_x}\left(\frac{\langle l_{e,K}\rangle}{L}\right)^{\frac{2[\alpha+1-2f(\alpha)]}{\alpha+3}} \mathrm{d}\alpha, \end{aligned} \tag{65}$$

where in this case α_x can be estimated from

$$\alpha_x(\xi) = \frac{2\ \ln\left[\dfrac{L}{\langle l_{e,K}\rangle}\right] - 3\ \ln\left[\dfrac{\xi^s_{\max}}{\xi}\right]}{2\ \ln\left[\dfrac{L}{\langle l_{e,K}\rangle}\right] - \ \ln\left[\dfrac{\xi^s_{\max}}{\xi}\right]} \tag{66}$$

in which $\xi^s_{\max}$ denotes the maximum particle diameter in the viscous subrange approximated by

$$\xi^s_{\max} = \frac{\sigma\nu_c^{1/2}}{\mu_c}\left[\frac{16(\mu_d/\mu_c)+16}{19(\mu_d/\mu_c)+16}\right]. \tag{67}$$

Baldyga and Podgórska (1998) developed two different breakage frequency models, using the multifractal formalism, to take into account the intermittent character of the fine-scale turbulence. For fluid particles with size falling into the inertial subrange, the breakage model was obtained considering the turbulent stresses as the main cause of breakage. On the other hand, for fluid particles with size less than Kolmogorov microscale length, the breakage mechanism can be considered viscous. Therefore, one of these models is suitable to describe turbulent or viscous breakage, respectively, in the inertial or viscous subrange. Furthermore, it is necessary point out that these models are reasonable for high Reynolds number flows but must be corrected for low Reynolds number flows.

4.1.6 Particle Breakage Model for Finite Reynolds Number

For low Reynolds number flows, the width of the inertial subrange becomes narrower, hence it may be necessary to extend the breakage models to the entire turbulence spectrum.

Solsvik and Jakobsen (2016) proposed a methodology to extend the Coulaloglou and Tavlarides (1977) and Luo and Svendsen (1996) models to the entire turbulence spectrum. They develop a new expression for the mean-squared velocity, $\bar{u}^2(\xi)$, from the energy spectrum model of Pope (2000):

$$\bar{u}^2(\xi) = \frac{4}{3}k\left(\frac{\xi^2}{(f(\mathrm{Re_L}))^2+\xi^2}\right)^{2/3} \times \{1-[T_1(\xi)+T_2(\xi)\cdot(T_3(\xi)T_4(\xi)-T_5(\xi))]\}, \tag{68}$$

where k is the turbulent kinetic energy, $f(\mathrm{Re_L})$ is a function of the Reynolds number, whereas $T_1(\xi)$, $T_2(\xi)$, $T_3(\xi)$, $T_4(\xi)$, and $T_5(\xi)$ are functions of the particle size (for their expressions refer to Solsvik and Jakobsen (2016)). Through Eq. (68) the mean turbulent kinetic energy, $\bar{E}$, in the Coulaloglou and Tavlarides (1977) model (Eq. 9) can be expressed as:

$$\bar{E} = c_{10}\rho_d\xi^3\frac{4}{3}k\left(\frac{\xi^2}{(f(\mathrm{Re_L}))^2+\xi^2}\right)^{2/3} \times \{1-[T_1(\xi)+T_2(\xi)\cdot(T_3(\xi)T_4(\xi)-T_5(\xi))]\}. \tag{69}$$

In the same way, using the Solsvik and Jakobsen (2016) expression for the mean-squared velocity, the Coulaloglou and Tavlarides (1977) breakage time (Eq. 11) can be rewritten as:

$$t_b = \frac{c_{11}\xi}{\left(\frac{4}{3}k\left(\frac{\xi^2}{(f(\mathrm{Re_L}))^2+\xi^2}\right)^{2/3}\{1-[T_1(\xi)+T_2(\xi)\cdot(T_3(\xi)T_4(\xi)-T_5(\xi))]\}\right)^{1/2}}. \tag{70}$$

Using Eqs. (69) and (70) in Eq. (6), the Coulaloglou and Tavlarides (1977) breakage frequency extended to entire turbulence spectrum was obtained. In a very similar way, the breakage model of Luo and Svendsen (1996) has also been extended by Solsvik and Jakobsen (2016) to entire turbulence spectrum.

Solsvik and Jakobsen (2016) point out that the extended breakage models depended of the Reynolds number, possibly accounting for many effects that result in numerous sets of empirical or semiempirical parameters present in other breakage frequency models. Hence, extending the models to the entire spectrum of turbulence might reduce experimental costs to refit the parameters.

4.2 Models for Other Breakage Mechanisms

As anticipated in Section 2, in some cases other breakage mechanisms different from turbulence fluctuations can become important and consequently appropriate models were developed for these conditions.

In systems in which macroscopic velocity gradients are important, deformation and fluid particle breakage depend on the capillary number (Eq. 4). In fact, it was shown that a fluid particle breaks if a capillary number reaches a critical value (Bentley, 1986; De Bruijn, 1993; Grace, 1982). Therefore, semiempirical model has been obtained in which the breakage time is a function of the capillary number (Elemans et al., 1993; Grace, 1982; Wieringa et al., 1996).

In gas–liquid flows when the balance between the viscous shear force and surface tension at the skirt of cap/slug fluid particles cannot be maintained, small fluid particles will be generated from the skirt of large fluid particles. This breakage mechanism, known as shearing-off process, depends from the relative velocity between the two phases. Fu and Ishii (2003a) developed a model in which the sheared-off volume is proportional to the relative velocity and the thickness of gas layer that may be sheared-off. Furthermore Fu and Ishii (2003a) speculated that the sheared-off particles size is proportional to the maximum stable particle diameter, which in turn can be estimated by We_{cr}.

Experimental evidences also show as fluid particles reach a limit of size, become unstable and disintegrate. This breakage mechanism was generally called surface instability. Wang et al. (2005a,b) proposed a breakage kernel due to this mechanism following the empirical correlation of Carrica and Clausse (1993). This model consists in a power law for the critical particles diameter, and assumes that the particle break in two equal daughters.

4.3 Breakage Kernel in Contaminated Systems

In the previous sections it was highlighted how the presence of surfactants, salts or more generally impurities might enhance breakage and at same time might inhibit coalescence, causing a decrease of the mean size of the disperse phase particles. Normally contamination effects, in the breakage kernels models described in the previous sections, are taken into account by simply using a different value for the surface tension, that is, considering its reduction due to surfactants present. Bak and Podgórska (2012), instead, in according with the observations of Koshy et al. (1988), modified their multifract breakage model (Baldyga and Podgórska, 1998), to consider the additional stress caused by the difference between dynamic surface tension and static surface tension. They obtained a new expression for the multifractal exponent α_x (Eq. 56), adding in the stress balance (Eq. 55) an extra disruptive stress generated by this difference in interfacial tension:

$$\tau_{dyn} \propto \frac{\sigma_0 - \sigma}{\xi}, \tag{71}$$

where σ_0 is the interfacial tension of the system without surfactants. Therefore, the modified multifractal exponent is given by (Bak and Podgórska, 2012):

$$\alpha_x(\xi) = \frac{\frac{5}{2}\ln\left[\frac{L\langle\epsilon\rangle^{0.4}\rho_c^{0.6}}{c_x(2\sigma-\sigma_0)^{0.6}}\right]}{\ln(L/\xi)} - 1.5. \tag{72}$$

However, in literature there are few models that take into account the breakage frequency rise, since usually the coalescence inhibition is the prevalent effect in the particles size reduction, therefore usually the contamination effect is taken into account in the coalescence kernel.

4.4 Comparison Among Different Breakage Kernel Models

Fig. 4 shows the trends of different breakage kernels as a function fluid particle size. A liquid–liquid system, i.e., water–toluene, and a gas–liquid system, i.e., air–water, were considered. The analysis was carried out for three different values of turbulent kinetic energy dissipation rate, ϵ, to evaluate the influence of this variable on the breakage kernels. In this regard,

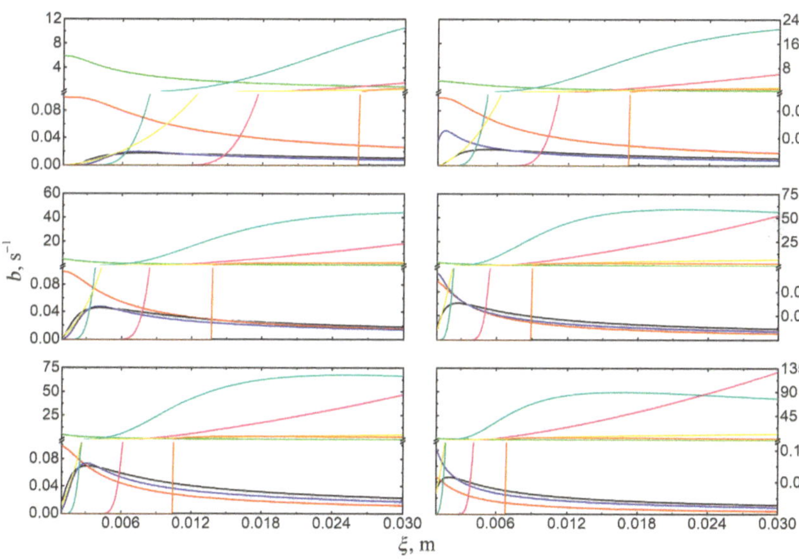

Fig. 4 Comparison of different breakage kernel models: Coulaloglou and Tavlarides (1977) (*black lines*), Narsimhan et al. (1979) (*red lines*), Alopaeus et al. (2002) (*green lines*), Baldyga and Podgórska (1998) (*blue lines*), Prince and Blanch (1990) (*cyan lines*), Lehr et al. (2002) (*magenta lines*), Luo and Svendsen (1996) (*yellow lines*), Martinez-Bazan et al. (1999a,b) (*orange lines*). The three graphs in the *left* refer to air–water system, whereas the three graphs in the *right* refer to toluene–water system. From *top* to *bottom* three different values for turbulent kinetic dissipation rate were used ($\epsilon = 0.01\ \mathrm{m^2\ s^{-3}}$, $\epsilon = 0.05\ \mathrm{m^2\ s^{-3}}$, $\epsilon = 0.1\ \mathrm{m^2\ s^{-3}}$).

from a preliminary analysis it can be noted that for both systems, all breakage models predict an increase of breakage frequency with the increase of turbulent kinetic energy dissipation rate. For the air–water system, the dependency over bubble size of the Coulaloglou and Tavlarides (1977) and Baldyga and Podgórska (1998) models is similar, in fact both kernels reach a maximum for a certain bubbles size, whose value depends on the magnitude of ϵ, with a subsequent monotonic decrease asymptotically tending to zero. For the toluene–water system, the plot shows how the maximum value is reached for particle size smaller than air–water system, because the air–water interfacial tension ($\sigma_{AW} = 0.073$ N m^{-1}) is greater than that for the toluene–water system ($\sigma_{TW} = 0.036$ N m^{-1}). Also the trends of Narsimhan et al. (1979) and Alopaeus et al. (2002) models show a maximum for small values of particle size and a subsequent decrease asymptotically to zero with the increase of particle sizes. Furthermore, with respect to the Coulaloglou and Tavlarides (1977) and Baldyga and Podgórska (1998) kernels, they are less sensitive to the value assumed by ϵ. It is interesting to note how the Alopaeus et al. (2002) model predicts a higher breakage frequency for the air–water system than for the toluene–water system, because it considers the restoring viscous stresses, which are greater in the latter due to the higher dynamic viscosity of toluene ($\mu_d = 5.83 \cdot 10^{-4}$ Pa s) than air ($\mu_d = 1.81 \times 10^{-5}$ Pa s). The Martinez-Bazan et al. (1999a,b) kernel is zero up to the critical size and after it increases rapidly, reaches a maximum (not visible in Fig. 4 due to the scale), depending on ϵ, and after it decreases slowly. Fig. 4 shows as the critical size decreases when the turbulent kinetic energy dissipation rate increases. Moreover, for equal values of ϵ, the critical size is smaller in the toluene–water system than in the air–water system, because the interfacial resistant stress, the only considered in this model, are greater in air–water system than in toluene–water. The Prince and Blanch (1990), Lehr et al. (2002), and Luo and Svendsen (1996) breakage kernels exhibit a growing trend with particle size (for Luo and Svendsen (1996) it is less visible in Fig. 4 due to scale). By comparing these models with those previously analyzed, it can be noted that for small values of particle size these models predict a lower breakage frequency than others, whereas for high values of particle size the opposite happens. Moreover, also in this case the magnitude of breakage frequency is higher in the toluene–water than in air water due to the different values of resistant interfacial stresses. However, the trend of Prince and Blanch (1990) kernel reaches a plateau for high values of the breakage frequency when the ϵ value is high enough. Anyway, the kernel in which the trends are growing monotonically could not be realistic, because the particles with size much larger than the characteristic scale of the inertial subrange could leave intact by eddies within the inertial subrange. Therefore, in this case the turbulent fluctuations breakage mechanism is ineffective and other breakage mechanisms should be considered. To conclude it is worth pointing out that all kernels analyzed depend on empirical parameters and in this investigation the values suggest in the original works (Alopaeus et al., 2002; Baldyga and Podgórska, 1998; Coulaloglou and Tavlarides, 1977; Lehr et al., 2002; Luo and Svendsen, 1996; Martinez-Bazan et al., 1999a,b; Narsimhan et al., 1979; Prince and Blanch, 1990) were employed.

5. MODELS FOR THE DAUGHTER SIZE DISTRIBUTION FUNCTION (DaSD)

The size of the daughter particles resulting from the breakage of a mother particle is a fundamental detail that is not directly provided by the breakage kernel described so far, with the exception of the models obtained through a partial breakage frequency (see Section 4.1). For this reason, appropriate models for the so-called daughter size distribution (DaSD) function have been proposed. Historically, the different models used to describe DaSD were classified in empirical, statistical, and phenomenological. The last two are more used than empirical approaches, because the latter can be employed only for the operative conditions for which the model parameters were determined. For this reason in this section only the statistical and phenomenological models will be briefly described.

5.1 Statistical Models

A statistical model assumes that the size of the daughter particles is a random variable and its probability distribution satisfies a certain statistical distribution, such as normal, beta, or uniform distribution functions. Valentas and Amundson (1966) assumed that the sizes of daughter particles are normally distributed about a mean value with a specific variance. Later the same approach was used by other authors, which speculated a truncated normal distribution function as DaSD (Chatzi et al., 1989; Coulaloglou and Tavlarides, 1977). Hsia and Tavlarides (1983) noted that the truncated normal function was unable to predict some experimental results (Ross et al., 1978; Verhoff et al., 1977), hence they used a beta function as DaSD, assuming binary breakage. Later, Lee et al. (1987a,b) modified this beta function, introducing two parameters empirically determined. Laakkonen et al. (2006, 2007) derived a beta function distribution in which a multiple breakage event is taken into account through an fitting parameter. Finally, other authors pointed out that a uniform distribution could be used because daughter particles of any size have an equal probability to form after a breakage event (Collins and Knudsen, 1970; Narsimhan et al., 1979; Prince and Blanch, 1990; Randolph, 1969). However, there is no physical argument to use a uniform model, because turbulent fluctuations are not uniform over all scales (Lasheras et al., 2002). On the other hand, both the normal and the beta distribution functions predict a maximum when the two daughter particles have the same size, which, for the gas–liquid system, is in disagreement with the experimental observations of Hesketh et al. (1991) and Andersson and Andersson (2006).

5.2 Phenomenological Models

A phenomenological model is not derived from theoretical fundamentals, but is simply constructed on empirical observations. The phenomenological models for DaSD are usually labeled according to the shape of the function: Bell-shaped, U-shaped and M-shaped. It is worth mentioning that all phenomenological models here described are based on the assumption of homogeneous and isotropic turbulence (Kolmogorov theory) and the size of the breaking particles lies within the inertial subrange of turbulence.

5.2.1 Bell-Shape

Martinez-Bazan et al. (1999b) derived a model for the DaSD on the basis of their experimental data analysis on air–water. In this model a parent particle, with size ξ, breaks into two daughter particles: the first one with size ξ_d, and the second one with complementary size $(\xi^3 - \xi_d^3)^{1/3}$. The fundamental hypothesis of the model is that the probability of finding a daughter particle with size in the range $\xi_{min} < \xi_d < \xi_{max}$ is proportional to the difference between the turbulent stresses over a length ξ_d and the resistant interfacial stresses of parent particle. Furthermore, supposing that the daughter size is a stochastic variable and uniformly distributed on the segment $[0, \xi]$, the DaSD can be expressed as (Martinez-Bazan et al., 1999b):

$$\beta(\xi, \xi_d^*) \propto \left[\frac{1}{2}\rho_c c_{12}(\epsilon \xi_d^* \xi)^{2/3} - \frac{6\sigma}{\xi}\right]\left[\frac{1}{2}\rho_c c_{13}\left(\epsilon\left(1 - \left(\xi_d^*/\xi\right)^3\right)\xi\right)^{2/3} - \frac{6\sigma}{\xi}\right], \tag{73}$$

where ξ is the parent particle size, $\xi_d^* = \xi_d/\xi$ is the dimensionless daughter size and $c_{12} = 8.2$. The above relation can be also expressed in dimensionless variables as (Martinez-Bazan et al., 1999b):

$$\beta(\Upsilon,\xi_{\rm d}^{*})\propto\left(\frac{1}{2}\rho_{\rm c}c_{12}(\epsilon\xi)^{2/3}\right)^{2}\left[(\xi_{\rm d}^{*})^{2/3}-\Upsilon^{5/3}\right]\left[(1-(\xi_{\rm d}^{*})^{3})^{2/9}-\Upsilon^{5/3}\right], \tag{74}$$

in which $\Upsilon = \xi_{\rm cr}/\xi$, and $\xi_{\rm cr}$ represents the size of the largest parent particle, which will not break due to the turbulent action of the flow. Finally, Martinez-Bazan et al. (1999b) obtained the final expression of DaSD using the normalization condition $\int_0^1 \beta(1,\xi_{\rm d}^{*})\mathrm{d}\xi_{\rm d}^{*}=1$, as:

$$\beta^{*}(1,\xi_{\rm d}^{*})=\frac{\left[(\xi_{\rm d}^{*})^{2/3}-\Upsilon^{5/3}\right]\left[(1-(\xi_{\rm d}^{*})^{3})^{2/9}-\Upsilon^{5/3}\right]}{\int_0^1\left[(\xi_{\rm d}^{*})^{2/3}-\Upsilon^{5/3}\right]\left[(1-(\xi_{\rm d}^{*})^{3})^{2/9}-\Upsilon^{5/3}\right]\mathrm{d}\xi_{\rm d}^{*}}. \tag{75}$$

After a detailed analysis, Zaccone et al. (2007) noticed the DaSD of Eq. (75) does not conserve volume. Hence, Martinez-Bazan et al. (2010) reviewed their previous model, pointing out that in the model derivation, the probability functions were erroneously expressed as a function of $\xi_{\rm d}^{*}$ rather than in term of daughter particle volume, $V_{\rm d}$. Hence the correct expression of Eq. (75) is (Martinez-Bazan et al., 2010):

$$\beta^{*}(1,V_{\rm d}^{*})=\frac{\left[(V_{\rm d}^{*})^{2/9}-\Upsilon^{5/3}\right]\left[(1-V_{\rm d}^{*})^{2/9}-\Upsilon^{5/3}\right]}{\int_0^1\left[(V_{\rm d}^{*})^{2/9}-\Upsilon^{5/3}\right]\left[(1-V_{\rm d}^{*})^{2/9}-\Upsilon^{5/3}\right]\mathrm{d}V_{\rm d}^{*}}, \tag{76}$$

where $V_{\rm d}^{*} = V_{\rm d}/V$, with V that represents the parent particle volume.

It can be easily verified that the bell-shape functions have the same behavior of some statistical models (normal and beta functions), that is, binary breakage is the most likely event and this is contrary to some experimental evidences (Andersson and Andersson, 2006; Hesketh et al., 1991).

5.2.2 U-Shape

Tsouris and Tavlarides (1994) proposed a model for the DaSD that consists in a bimodal function with the highest probability density in finding smallest and largest daughter particles and lower probability density two daughters with the same size. However, Tsouris and Tavlarides (1994) postulated that the DSaD is linearly related to energy requirement for the formation of the daughter particles, expressing the normalized DSD function as:

$$\beta^*(\xi,\xi_d)=\frac{\Delta E_{min}+\Delta E_{max}-\Delta E\left(\xi_d,\left(\xi^3-\xi_d^3\right)^{1/3}\right)}{\int_0^{\xi}\left[\Delta E_{min}+\Delta E_{max}-\Delta E\left(\xi_d,\left(\xi^3-\xi_d^3\right)^{1/3}\right)\right]d\xi_d}, \tag{77}$$

in which E_{min} denotes the energy needed to create the smallest and largest daughter particles, E_{max} is the energy required to create two equal-size daughter particles, whereas $E\left(\xi_d,\left(\xi^3-\xi_d^3\right)^{1/3}\right)$ is the energy needed to form two daughter particles with size ξ_d and $\left(\xi^3-\xi_d^3\right)^{1/3}$, respectively. It is important to highlight that, as in the statistical models, this model cannot describe a physical situation because it is independent from the parent size and flow conditions (Liao and Lucas, 2009).

Another U-shape distribution function was obtained by Luo and Svendsen (1996) through the ratio between the partial and the overall breakage frequencies (see Section 4.1.3). Differently by the Tsouris and Tavlarides (1994) model, the DaSD function obtained by Luo and Svendsen (1996) has a minimum equally sized daughters that is not zero, but it depends from the parent particle size.

5.2.3 M-Shape

Lehr et al. (2002), as Luo and Svendsen (1996), derived the DaSD from the ratio between partial and overall breakage frequencies (see Section 4.1.3). The resulting DaSD has the following expression:

$$\beta(\xi,\xi_d)=\frac{9\xi_d^2}{\sqrt{\pi}\xi^3}\frac{\exp\left\{\frac{9}{4}\left[\ln\left(\frac{2^{2/5}\xi_d\rho_c^{3/5}\epsilon^{2/5}}{\sigma^{3/5}}\right)\right]^2\right\}}{1+\operatorname{erfc}\left[\frac{3}{2}\ln\left(\frac{2^{15}\xi\rho_c^{3/5}\epsilon^{2/5}}{\sigma^{3/5}}\right)\right]}. \tag{78}$$

According to this model the probability of finding small and large daughter particles increases with the parent size, however, the position of the peak does not change much. Furthermore, with the increase of parent size, the distribution change from monomodal to bimodal.

Using the same conceptual approach of Luo and Svendsen (1996) and Lehr et al. (2002), also Wang et al. (2003) and Zhao and Ge (2007) obtained an M-shape DaSD function. In particular the model proposed by Zhao and Ge (2007) is slightly different from the others, since they assumed that in binary breakage, one of the daughter particles has the size of the interacting eddy. Therefore, the probability density of a daughter particle with volume, V_{d}, equals the sum of the breakage induced by all eddies with size $l_e = (6V_{\mathrm{d}}/\pi)^{1/3}$ and size $l_e = (6(V - V_{\mathrm{d}})/\pi)^{1/3}$:

$$\beta^*(\xi, \xi_{\mathrm{d}}) = \frac{w(\xi, l_e)P_{\mathrm{b}}(\xi, l_e)}{\int_{l_{e,\min}}^{\xi} w(\xi, l_e)P_{\mathrm{b}}(\xi, l_e)\mathrm{d}\xi} + \frac{w(\xi, \xi - l_e)P_{\mathrm{b}}(\xi, \xi - l_e)}{\int_{l_{e,\min}}^{\xi} w(\xi, \xi - l_e)P_{\mathrm{b}}(\xi, \xi - l_e)\mathrm{d}\xi}, \tag{79}$$

where $w(\xi, l_e)$ and $P_{\mathrm{b}}(\xi, l_e)$ are, respectively, the collision frequency and the collision efficiency (see Section 4.1.3).

At the moment, there is not a general consensus in the literature regarding which DaSD form, bell-shape, U-shape or M-shape, should be used. The reason is that it is difficult to perform experiments and correlate the experimental results in the form of a DaSD. The validation of DaSD is often indirect, through the solution of a PBE and comparison of the predicted and experimental PSD, rather than a direct verification of the number of particles formed according to the parent particle size and flow condition. Therefore in the future investigation are required on this aspect. Moreover, it is worth mentioning that it is not know whether the presence of impurities has an influence on the number of particles formed after the breakage and therefore it is neglected in the formulation of DaSD for such systems.

6. ANALYSIS OF DIFFERENT MODELS FOR FLUID PARTICLE COALESCENCE KERNEL

Empirical and physical considerations were used to derive the expression for coalescence kernel. In general, the empirical models were developed to correlate the experimental data obtained in stirred tanks through power or exponential laws depending on fluid particles size, disperse phase volume fraction, agitation speed, and empirical parameters (Casamatta and Vogelpohl, 1985; Kentish et al., 1998; Konno et al., 1988; Wright and Ramkrishna, 1994). As in all empirical models, they depend on the experimental setup or geometrical parameters and so these models cannot generalized to other cases. On the other hand, physical models are more used than empirical ones and a generic physical model is usually expressed as:

$$a(\xi_1,\xi_2) = h(\xi_1,\xi_2)\eta(\xi_1,\xi_2), \tag{80}$$

where $h(\xi_1, \xi_2)$ denotes the collision frequency, whereas $\eta(\xi_1, \xi_2)$ represents the coalescence efficiency.

6.1 Models for Collision Frequency $h(\xi_1, \xi_2)$

As mentioned in Section 3, different mechanisms promote collision between fluid particles. Usually, the main collision mechanism in turbulent systems is the one caused by turbulent velocity fluctuation, whereas other mechanisms play a marginal role. Therefore, only the models obtained for turbulent velocity fluctuations will be analyzed in details and only a brief description will be devoted to the other ones.

6.1.1 Collisions Due to Turbulent Velocity Fluctuations

As in some breakage kernels (Luo and Svendsen, 1996; Prince and Blanch, 1990; Tsouris and Tavlarides, 1994), the collisions among fluid particles induced by turbulent velocity fluctuations are usually assumed to be similar to collisions between gas molecules in gas kinetic theory. Therefore, the collision frequency between two fluid particles, with size ξ_1 and ξ_2, can be expressed as the volume swept by the moving particle in the unit time (Kennard, 1938):

$$h(\xi_1,\xi_2) = S(\xi_1,\xi_2)u_{\mathrm{rel}}(\xi_1,\xi_2), \tag{81}$$

where $S(\xi_1, \xi_2)$ is the collision-sectional area and $u_{\mathrm{rel}}(\xi_1, \xi_2)$ is the corresponding relative velocity. Assuming spherical fluid particles, the collision-sectional area can be easily estimated as:

$$S(\xi_1,\xi_2) = \frac{\pi}{4}(\xi_1 + \xi_2)^2. \tag{82}$$

To estimate the particle relative velocity u_{rel}, normally it is assumed that the particle velocity is equal to the eddy velocity with the same size of the particles (Coulaloglou and Tavlarides, 1977; Lee et al., 1987a; Luo, 1993; Prince and Blanch, 1990) and that a small eddy does not have a sufficient energy to influence particles motion. It is also assumed that much larger eddy only transports the fluid particles without affecting the relative velocity. From these hypothesis the particle relative velocity can be expressed as:

$$u_{\mathrm{rel}} = \sqrt{\bar{u}^2(\xi_1) + \bar{u}^2(\xi_2)}. \tag{83}$$

Supposing isotropic and homogeneous turbulence and the particle size falling in the inertial subrange, the velocities $\bar{u}(\xi_1)$ and $\bar{u}(\xi_2)$ can be expressed using classical turbulence theory (Kolmogorov, 1941) through Eq. (10). On the other hand, Chesters (1991) employed a different expression for the particles relative velocity:

$$u_{\mathrm{rel}} = \epsilon^{1/3}(\xi_1 + \xi_2)^{1/3}. \tag{84}$$

The resulting collision frequency obtained by replacing of Eqs. (82) and (83) into Eq. (81) (Coulaloglou and Tavlarides, 1977; Lee et al., 1987a; Luo, 1993; Prince and Blanch, 1990) reads as follows:

$$h(\xi_1, \xi_2) = c_{15}\frac{\pi}{4}(\xi_1 + \xi_2)^2\left(\xi_1^{2/3} + \xi_2^{2/3}\right)^{1/2}\epsilon^{1/3}, \tag{85}$$

where c_{15} is an empirical parameter. In subsequent works some modifications to Eq. (85) that consist in multiplying the collision frequency by two factors (Θ and Λ) were proposed. The Θ factor takes into account the possible presence of other fluid particles among the two colliding, with the consequent rise in collision frequency. Wu et al. (1998) proposed the following expression:

$$\Theta = \frac{1}{\alpha_{\mathrm{d,max}}^{1/3}\left(\alpha_{\mathrm{d,max}}^{1/3} - \alpha_{\mathrm{d}}^{1/3}\right)}, \tag{86}$$

where $\alpha_{\mathrm{d,max}}$ is the maximum possible dispersed phase holdup, fixed to 0.8. Similar expressions for Θ factor were subsequently proposed (Hibiki and Ishii, 2000a,b; Wang et al., 2005a,b), while Lehr et al. (2002) used the following function:

$$\Theta = \exp\left[-\left(\frac{\alpha_{\mathrm{d,max}}^{1/3} - \alpha_{\mathrm{d}^{1/3}}}{\alpha_{\mathrm{d}}}\right)^2\right], \tag{87}$$

where $\alpha_{\mathrm{d,max}} = 0.6$. The Λ factor is a function of the ratio between the mean distance between particles and their relative turbulent path length. In fact, if the distance between particles is larger than the turbulent path length, the collision should not be counted in the collision frequency (Wang et al., 2005a,b; Wu et al., 1998). Λ can be considered as a damping factor collision frequency. Wu et al. (1998) proposed the following expression:

$$\Lambda = \left[1 - \exp\left(-c_{16}\frac{l_e}{H_{1,2}}\right)\right], \tag{88}$$

where l_e is average size of the eddy that drives particles of size ξ_1 and ξ_2 together, and $H_{1,2}$ represents the mean distance between the particles. From Eq. (88) it can be noted that if $l_e/H_{1,2}$ is small, Λ tends to zero. On the other hand, Wang et al. (2005a,b) employed a different relation for Λ:

$$\Lambda = \exp\left[-\left(\frac{H_{1,2}}{l_{1,2}}\right)^6\right], \tag{89}$$

where, in this case, $l_{1,2}$ is the mean relative turbulent path length of the particles with size ξ_1 and ξ_2. The modified form of the collision frequency proposed by Wang et al. (2005a,b) can be rewritten as:

$$h(\xi_1, \xi_2) = c_{16}\Theta\Lambda\frac{\pi}{4}(\xi_1 + \xi_2)^2(\xi_1^{2/3} + \xi_2^{2/3})^{1/2}\epsilon^{1/3}. \tag{90}$$

From Eq. (90) it can easily be noted that due to the Λ factor, the collision frequency for small particles is generally lower than for large particles, because the mean distance between the small particles is usually larger than between big ones (Liao and Lucas, 2010).

The multifractal approach, briefly described in Section 4.1.5, was also extended to consider particle coalescence. In fact, through the multifractal description an expression for the collision frequency was derived, by using the characteristic velocity in intermittent turbulent flow (Eq. 49) to describe the relative velocity among colliding particles (Podgórska and Baldyga, 2001):

$$h(\xi_1, \xi_2, \alpha) = \sqrt{\frac{8\pi}{3}}\epsilon^{1/3}\left(\frac{\xi_1 + \xi_2}{2}\right)^{7/3}\left(\frac{\xi_1 + \xi_2}{2L}\right)^{(\alpha-1)/3}, \tag{91}$$

where also in this case α and L represent, respectively, the multifractal exponent and the integral scale of turbulence. By taking into account the probability density for α and employing Eq. (53) to approximate the multifractal spectrum, the following expression was derived (Podgórska and Baldyga, 2001):

$$h(\xi_1,\xi_2) = \sqrt{\frac{8\pi}{3}}\epsilon^{1/3}\left(\frac{\xi_1+\xi_2}{2}\right)^{7/3}\left(\frac{\xi_1+\xi_2}{2L}\right)^{0.026}. \tag{92}$$

Eq. (92) depends of the integral scale of turbulence, L, and therefore accounts for the influence of the system scale. In fact, it can be noted that intermittency effects lower the collision rate only in large systems, where $(\xi_1 + \xi_2)/L \ll 1$.

As for the breakage kernel, Solsvik and Jakobsen (2016) proposed to extend the collision frequency to the entire turbulence spectrum, in order to obtain a coalescence kernel not limited only to the inertial subrange. The Solsvik and Jakobsen (2016) approach consists in estimating the relative velocity using Eq. (68) rather than Eq. (10), resulting in:

$$\begin{aligned} h(\xi_1,\xi_2) &= c_{17}(\xi_1+\xi_2)^2 \\ &\times\Bigg[\left(k\left(\frac{\xi_1^2}{(f(\mathrm{Re_L}))^2+\xi_1^2}\right)^{2/3}\{1-[T_1(\xi_1)+T_2(\xi_1)\cdot(T_3(\xi_1)T_4(\xi_1)-T_5(\xi_1))]\}\right) \\ &+\left(k\left(\frac{\xi_2^2}{(f(\mathrm{Re_L}))^2+\xi_2^2}\right)^{2/3}\{1-[T_1(\xi_2)+T_2(\xi_2)\cdot(T_3(\xi_2)T_4(\xi_2)-T_5(\xi_2))]\}\right)\Bigg]^{1/2}. \end{aligned} \tag{93}$$

where $T_1(\xi_2)$, $T_2(\xi_2)$, $T_3(\xi_2)$, $T_4(\xi_2)$, and $T_5(\xi_2)$ are the same function employed in Eq. (68).

6.1.2 Collisions Due to Other Mechanisms

The other collision mechanisms described in Section 3 can be usually neglected in the estimation of the collision rate for turbulent systems, but in particular situations they may become important.

Friedlander (1977) proposed a model to consider the collisions caused by macroscopic shear rate in a uniform laminar flow, assuming that the streamlines are straight and particle motion rectilinear. The same model was employed by Tavlarides and Stamatoudis (1981) and Prince and Blanch (1990).

When the particle size is very small and falls in the dissipation subrange, the forces that govern collisions are mainly viscous. In this case, the collision frequency should be determined by the local shear of the flow in turbulent eddies and its form is similar to the shear-induced collision model proposed by Friedlander (1977) (Chesters, 1991).

In the past, collisions due to different rising velocities of the fluid particles were considered in many models. These models have the same form of Eq. (81) with the difference that the relative velocity is computed from buoyancy force (Friedlander, 1977; Liao et al., 2015; Prince and Blanch, 1990; Wang et al., 2005a,b).

In the last decades, it was experimentally observed that in gas–liquid systems the wake generated by rising large fluid particles can play a role in the collision between the particles especially in fluids sufficiently viscous. Unlike the other collision mechanisms, different approaches were used to develop a model to account for this collisions, see, for example, the works of Kalkach-Navarro et al. (1994), Wu et al. (1998), Hibiki and Ishii (2000a,b), Hibiki et al. (2001), Fu and Ishii (2003a,b), and Wang et al. (2005a,b).

6.2 Models for Coalescence Efficiency $\eta(\xi_1, \xi_2)$

In the coalescence kernel model, an efficiency term is fundamental to count only the effective collisions. As anticipated in the Section 3, three different approaches were developed to obtain the coalescence efficiency: energy model, critical approach velocity model, and film drainage model.

6.2.1 Energy Model

Howarth (1967) first noted that only a fraction of the collisions resulted in immediate coalescence and that the coalescence probability increased with the collision energy. This observation was subsequently confirmed by some experiments (Kuboi et al., 1972; Park and Blair, 1975), and on the basis of these, Sovová (1981) proposed a model in which the probability of collision leading to coalescence is determined by the ratio between the interfacial energy, E_σ, and the collision energy, E_{kin}:

$$\eta(\xi_1, \xi_2) = \exp\left(-c_{18}\frac{E_\sigma}{E_{kin}}\right). \tag{94}$$

The interfacial energy was expressed as proportional to the interfacial tension and the colliding particles surface area:

$$E_\sigma \propto \sigma\left(V_1^{2/3} + V_2^{2/3}\right), \tag{95}$$

while the collision kinetic was defined as,

$$E_{kin} = \frac{1}{2}\rho_d \frac{V_1 V_2}{V_1 + V_2} u_{rel}^2, \tag{96}$$

where u_{rel} is the relative velocity estimate through Eqs. (83) and (10), resulting in (Sovová, 1981):

$$\eta(\xi_1, \xi_2) = \exp\left[-c_{19} \frac{\sigma(V_1^{2/3} + V_2^{2/3})(V_1 + V_2)}{\rho_d \epsilon^{2/3} V_1 V_2 (V_1^{2/9} + V_2^{2/9})}\right]. \tag{97}$$

Another energy model was proposed by Simon (2004) that estimated the collision kinetic energy from momentum balance during collision:

$$E_{kin} \propto \rho_d \epsilon^{2/3} (V_1^{11/9} + V_2^{11/9}), \tag{98}$$

resulting in:

$$\eta(\xi_1, \xi_2) = \exp\left[-c_{20} \frac{\sigma(V_1^{2/3} + V_2^{2/3})}{\rho_d \epsilon^{2/3} (V_1^{11/9} + V_2^{11/9})}\right]. \tag{99}$$

6.2.2 Critical Approach Velocity Model

Doubliez (1991) and Duineveld (1994) observed experimentally that in a gas–liquid system, small fluid particles and small approach velocities lead to a higher coalescence efficiency. About a decade later, Lehr et al. (2002) through air–water experiments estimated the maximum value of the approach velocity, known as critical velocity, resulting in coalescence, $u_{cr} = 0.08$ m/s. The Lehr et al. (2002) efficiency model states that:

$$\eta(\xi_1, \xi_2) = \begin{cases} 1 & u_{rel} \le u_{cr}; \\ 0 & u_{rel} > u_{cr}. \end{cases} \tag{100}$$

This approach can be considered empirical, since it is necessary to determine experimentally the critical approach velocity for each investigated system. For example, Lehr et al. (2002) pointed out that $u_{cr} = 0.08$ m/s was obtained for distillate water and could decrease with the system contamination.

6.2.3 Film Drainage Model

As discussed in Section 3, this model determines the coalescence efficiency through two characteristic time scales: the contact time, i.e., the time for which the two particles staying in contact after the collision, and the drainage time, i.e., the time required by the fluid between two particles to drain out. Various regimes of drainage may be distinguished, depending on the rigidity and mobility of the fluid particles interfaces. In fact, very small particles can be considered as rigid spheres (see Fig. 5), whereas contaminants play a fundamental role in the interface mobility (see Fig. 6). The drainage time is generally estimated by the integration of the thinning equation, which in turn is obtained through the lubrication theory, i.e., solving the Navier–Stokes and the continuity equation for simple boundary conditions.

6.2.3.1 Nondeformable Particles With Immobile Interfaces

According to Chesters (1991), the mean radial velocity in the film implies (via the Poiseuille relation for flow between quasi-parallel surfaces) a certain pressure variation and hence a certain interaction force:

$$F=\frac{3\pi\mu_c r^2}{2g}\frac{dg}{dt}, \tag{101}$$

where g is the distance between the interfaces of the fluid particles, r is the particles radius, F is the compression force and dg/dt is the approach velocity. If F is constant and integrating Eq. (101) from g_i to g_f, the following expression is obtained (Chesters, 1991):

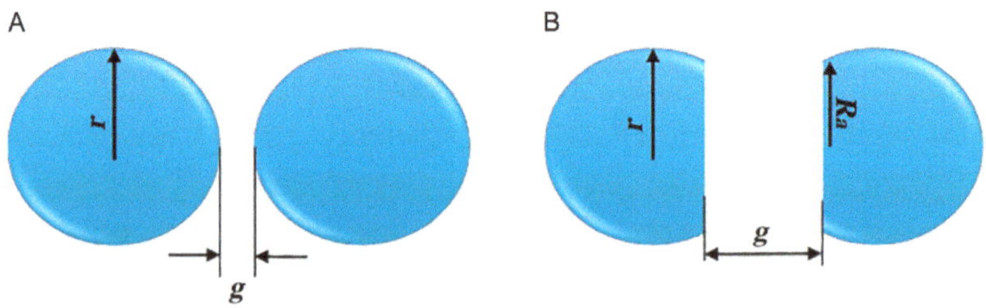

Fig. 5 (A) Nondeformable particles; (B) Deformable particles.

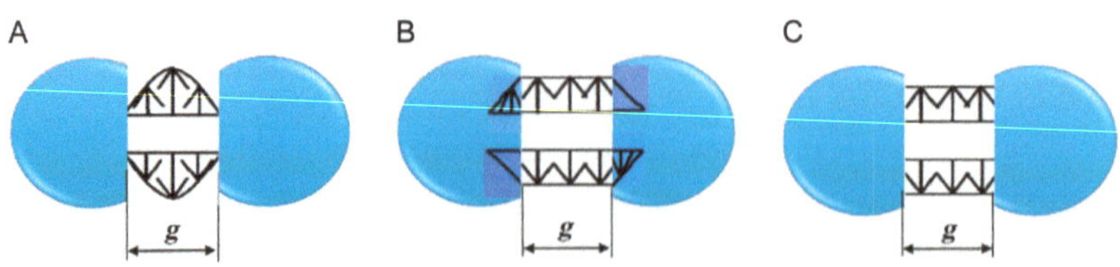

Fig. 6 (A) Deformable particle with immobile interfaces; (B) Deformable particles with partial mobile interfaces; (C) Deformable particles with fully mobile interfaces.

$$t_{\text{drainage}} = \frac{3\pi\mu_c}{2F} r^2 \ln\left(\frac{g_i}{g_f}\right). \tag{102}$$

Eq. (102) was derived for two equal nondeformable fluid particles, but it can be extended to two unequal particles by replacing r with the equivalent radius: $r_{eq} = 2r_1r_2/(r_1 + r_2)$. Usually, in turbulent systems it is assumed that the compression force is generated by turbulent eddies, which pull the fluid particles close to each other. Therefore, F is assumed proportional to the mean-squared velocity difference at either ends of the eddy with size of $l_e = \xi_1 + \xi_2$ (Coulaloglou and Tavlarides, 1977; Tsouris and Tavlarides, 1994):

$$F \propto \rho_c \bar{u}^2 \left(\frac{\xi_1\xi_2}{\xi_1 + \xi_2}\right)^2. \tag{103}$$

Furthermore, if the eddies responsible for collisions belong to the inertial subrange, $\bar{u}$ can be expressed through Eq. (10) resulting in:

$$F \propto \rho_c \epsilon^{2/3} (\xi_1 + \xi_2)^{2/3} \left(\frac{\xi_1\xi_2}{\xi_1 + \xi_2}\right)^2. \tag{104}$$

On the other hand, Podgórska (2005) proposed an expression for the compression force to account for the turbulence intermittency:

$$F \propto r_{eq}^2 \rho_c \epsilon^{2/3} \left(\frac{\xi_1 + \xi_2}{2}\right)^{2/3} \left(\frac{\xi_1 + \xi_2}{2L}\right)^{(2/3)(\alpha-1)}. \tag{105}$$

Eq. (105) was employed in the following thinning rate equation to obtain the drainage time for two spherical particles (Davis et al., 1989):

$$F = \frac{3\pi\mu_c}{2g} r_{eq}^2 \frac{dg}{dt} \frac{1 + 0.38\,\mathrm{m}}{1 + 1.69\,\mathrm{m} + 0.43\,\mathrm{m}^2}, \tag{106}$$

where $m = (\mu_c/\mu_d)\sqrt{r_{eq}/h}$ (in this case set to 0) takes into account the eventual mobility of the interface. By integrating Eq. (106) from the initial film thickness, g_i, to the critical film thickness g_f and applying the steepest descent method, the following drainage time was derived (Podgórska, 2005):

$$t_{\text{drainage}} = \frac{3\pi\mu_c}{2\rho_c \epsilon^{2/3}} \left(\frac{\xi_1 + \xi_2}{2}\right)^{-2/3} \left(\frac{2L}{\xi_1 + \xi_2}\right)^{0.13} G_q, \tag{107}$$

where G_q is a logarithmic function of g_i and g_f.

The assumption of a nondeformable particles is reasonable when the disperse phase viscosity is much larger than the continuous phase viscosity or for very small particles ($\xi < 1$ mm). In all other cases the interface deformation must be considered (Simon, 2004).

6.2.3.2 Deformable Particles With Immobile Interfaces

As mentioned large particles are generally assumed deformable whereas immobile interfaces usually (Fig. 6A) refer to contaminated systems (e.g., air–water dispersions with high surfactant concentrations). In these systems, it is assumed that the continuous phase is drained from the rigid interfaces as a laminar flow with a parabolic velocity profile. Chesters (1991) obtained the following expression for the approach velocity:

$$-\frac{\mathrm{d}g}{\mathrm{d}t}=\frac{2g^3F}{3\pi\mu_c R_a^4}, \tag{108}$$

where in this case the interaction force F was assumed proportional to the excess pressure in the film:

$$F \propto \pi R_a^2 \frac{2\sigma}{r_{eq}}. \tag{109}$$

Integrating Eq. (108) from the distance which the collision occurs, g_i, to the rupture film distance, g_f, the drainage time can be estimated as follows (Chesters, 1991):

$$t_{drainage}=\frac{3\mu_c F}{16\pi\sigma^2}\left(\frac{\xi_1\xi_2}{\xi_1+\xi_2}\right)^2\left(\frac{1}{g_f^2}-\frac{1}{g_i^2}\right). \tag{110}$$

Eq. (110) is the same obtained by Coulaloglou and Tavlarides (1977). They employed Eq. (104) for F:

$$t_{drainage}=c_{21}\frac{3\mu_c\rho_c}{16\pi\sigma^2}\epsilon^{2/3}(\xi_1+\xi_2)^{2/3}\left(\frac{\xi_1\xi_2}{\xi_1+\xi_2}\right)^4, \tag{111}$$

where c_{21} is a dimensional system-dependent constant ($1/\mathrm{m}^2$) that includes g_i and g_f. Podgórska (2005), modified the Chesters (1991) model to account for turbulence intermittency, resulting in (Podgórska, 2005):

$$t_{drainage}=\frac{3}{16\sigma^2 g_f^2}\mu_c\rho_c r_{eq}^4\epsilon^{2/3}\left(\frac{\xi_1+\xi_2}{2}\right)^{2/3}\left(\frac{\xi_1+\xi_2}{2L}\right)^{0.026}. \tag{112}$$

6.2.3.3 Deformable Particles With Partially Mobile Interfaces

This condition applies to systems with a certain degree of contamination. By assuming the quasi-steady creeping flow, Chesters (1991) obtained a relation for the shear stress in the particle interface as a function of the surface velocity which can be considered the film velocity, and considering the pressure variation in the film:

$$t_{\text{drainage}} = \frac{\pi \mu_{\text{d}} F^{1/2}}{2(2\pi\sigma/r_{\text{eq}})^{3/2}} \left(\frac{1}{g_f} - \frac{1}{g_i} \right) \tag{113}$$

where F can be estimated through Eq. (109). Podgórska and Baldyga (2001) modified the above model, using an expression for g_i estimated by comparing the characteristic turbulent velocity expressed in Eq. (49) resulting in:

$$t_{\text{drainage}} = \frac{\mu_{\text{d}} \widetilde{R}_a r_{\text{eq}}^{3/2}}{4\sigma\xi_1^{1/2}} \left[\frac{1}{g_f} \left(\frac{\xi_1 + \xi_2}{2L} \right)^{0.016} - \frac{1}{\widetilde{g}_i} \left(\frac{\xi_1 + \xi_2}{2L} \right)^{-0.01} \right], \tag{114}$$

where $\widetilde{R}_a$ and $\widetilde{g}_i$ are the film radius and the initial thickness, respectively, estimated when neglecting intermittency. They are related to film radius $R_a = \widetilde{R}_a ((\xi_1 + \xi_2)/2L)^{(\alpha-1)/6}$ and initial film thickness $g_i = \widetilde{g}_i ((\xi_1 + \xi_2)/2L)^{(\alpha-1)/4}$ in intermittent turbulence. Lee et al. (1987a) used the model proposed by Sagert and Quinn (1976) for the partially mobile interfaces:

$$t_{\text{drainage}} = -3M\mu_{\text{c}} R_a^2 \int_{g_i}^{g_f} \frac{\mathrm{d}g}{8g^3[(2\sigma)/r + A_{\text{h}}/(6\pi g^3)]}, \tag{115}$$

where A_{h} is the Hamaker constant and M is the surface immobility parameter which assumes the value 1, 2, or 4 depending on whether there are 0, 1, or 2 immobile interface, respectively. Tsouris and Tavlarides (1994) proposed a drainage time model for partially mobile interface using Eqs. (106) and (104).

6.2.3.4 Deformable Particles With Fully Mobile Interfaces

In systems in which the contamination is very low and the disperse phase viscosity is low, fluid particle interfaces can be considered fully mobile (Fig. 6C). Chesters (1975) used the parallel film model to derive a drainage velocity model which includes the contribution of inertia and viscous forces:

$$\frac{\mathrm{d}G}{\mathrm{d}t}=\left[\frac{\sigma}{3\mu_c r}\left(\frac{\mathrm{d}G}{\mathrm{d}t}\right)_0\right]\exp\left(-\frac{12\mu_c t}{\rho_c R_a^2}\right)-\frac{\sigma}{3\mu_c r}, \tag{116}$$

where $G=1/2\ \ln g$. For this general equation there is no analytical solution, therefore two limiting cases are considered. For highly viscous liquids the viscous term is prevalent and Eq. (116) becomes:

$$\frac{\mathrm{d}g}{\mathrm{d}t}=-\frac{2g\sigma}{3\mu_c r}, \tag{117}$$

which effectively represents the viscous resistance to squeezing between two perfectly lubricated slabs. Furthermore, Eq. (116) is independent of the film dimension, R_a, and hence of the applied force, F. Finally, for this case the drainage time can be obtained from integration Eq. (117):

$$t_{\mathrm{drainage}}=\frac{3\mu_c r}{2\sigma}\ln\left(\frac{g_i}{g_f}\right). \tag{118}$$

In the inertia-controlled limit, Eq. (116) admits the following solution for the drainage time:

$$t_{\mathrm{drainage}}=\frac{\rho_c r^2 u_{\mathrm{rel}}}{8\sigma}\ln\left(\frac{g_i}{g_f}\right), \tag{119}$$

where u_{rel} is the relative velocity of the colliding particles and r is their radius in the case in which they have the same size. This model can be extended to unequal particles replacing r with r_{eq} (Luo, 1993). Eq. (119) is proportional to particle relative velocity and hence the gentle collisions increase the coalescence probability according to the experimental observations of Lehr et al. (2002). For the coalescence of two equally sized fluid particle in a pure liquid with low viscosity ($\mu_c \leq 10$ mPa s) Lee et al. (1987a) employed the following relation for drainage time (Sagert and Quinn, 1976):

$$t_{\mathrm{drainage}}=\frac{R_a}{4}\left(\frac{\rho_c \xi}{2\sigma}\right)^2\ln\left(\frac{g_i}{g_f}\right). \tag{120}$$

Prince and Blanch (1990) derived an expression from the following thinning rate relation (Oolman and Blanch, 1986):

$$-\frac{\mathrm{d}g}{\mathrm{d}t}=\left\{\frac{8}{R_a^2\rho_c}\left[-\frac{4C_{sur}}{R_g T}\left(\frac{\mathrm{d}\sigma}{\mathrm{d}c}\right)^2+g^2\left(\frac{2\sigma}{r}+\frac{A_{\mathrm{h}}}{6\pi g^3}\right)\right]\right\}^{1/2}, \tag{121}$$

where R_g is the gas constant, T is the temperature, A_h is the Hamaker constant, and C_{sur} is the concentration of surfactant species. Prince and Blanch (1990) applied this model for distilled water case, hence considered the concentration of contaminants negligible. Furthermore, they highlighted how the effect of the Hamaker force becomes appreciable at very low film thicknesses, just prior to rupture. Therefore, neglecting the Hamaker force contribution and for $C_{sur} = 0$, the following relation was obtained by integration of Eq (121):

$$t_{\text{drainage}} = \left(\frac{r_{\text{eq}}\rho_{\text{c}}}{16\sigma}\right)^{1/2} \ln\left(\frac{g_i}{g_f}\right). \tag{122}$$

The result of neglecting the Hamaker force contribution may lead to a slight overestimation of the coalescence time, but on the other hand the numerical solution for Eq. (121) is no more necessary.

The most simple relation to estimate contact time was proposed by Levich (1962) through a dimensional analysis:

$$t_{\text{contact}} \propto \frac{\xi^{2/3}}{\epsilon^{1/3}}. \tag{123}$$

This relation was employed by Coulaloglou and Tavlarides (1977) and other authors to develop they respective models. Also Podgórska (2005) estimated an expression for the contact time through a dimensional analysis, using the multifractal formalism:

$$t_{\text{contact}} = \frac{\xi_{1,2}^{2/3}}{\langle\epsilon\rangle^{1/3}} \left(\frac{\xi_{1,2}}{L}\right)^{(1-\alpha)/3}, \tag{124}$$

where $\xi_{1,\,2} = (\xi_1 + \xi_2)/2$. Chesters (1991) pointed out that in the collision the particle kinetic energy is converted into surface energy during the film formation process:

$$8\pi r^2 \sigma \left(\frac{a}{2R}\right)^4 = \frac{1}{2}\pi r^3 \left(\frac{4\rho_{\text{d}}}{3} + C_{\text{vm}}\rho_{\text{c}}\right) u_{\text{rel}}^2, \tag{125}$$

where C_{vm} is the virtual mass coefficient. It can be easily show how the distance from the onset of flatting up to the point at which the particle motion is arrested is of order of $(R_a/r)^2$. Hence, after obtaining this distance from Eq. (125), the expression of the contact time reads as follows:

$$t_{\text{conctact}} \approx \frac{(R_a/r)^2}{u_{\text{rel}}} = \left[\frac{(4\rho_{\text{d}}/3\rho_{\text{c}} + C_{\text{vm}})\rho_{\text{c}} r^3}{2\sigma}\right]^{1/2}. \tag{126}$$

Using similar considerations, this model was subsequently extended to unequally sized particles by Kamp et al. (2001):

$$t_{\text{conctact}} = \frac{\pi}{4}\left(\frac{\rho_{\text{c}} C_{\text{vm}} \xi_{\text{eq}}^3}{3\sigma}\right)^{1/2}, \tag{127}$$

where $\xi_{\text{eq}} = (2\xi_1\xi_2)/(\xi_1 + \xi_2)$, and C_{vm} is usually fixed constant between 0.5 and 0.8. Another similar model, for a simple parallel film, was derived by Luo (1993):

$$t_{\text{conctact}} = (1 + \xi_1/\xi_2)\left(\frac{(\rho_{\text{d}}/\rho_{\text{d}} + C_{\text{vm}})\rho_{\text{c}} \xi_1^3}{3(1 + (\xi_1/\xi_2)^2)(1 + (\xi_1/\xi_2)^3)\sigma}\right)^{1/2}. \tag{128}$$

In addition to these theoretical models, semiempirical models were also used to estimate the contact time. For instance, Tsouris and Tavlarides (1994) employed the following expression obtain for stirred tanks (Schwartzberg and Treybal, 1968):

$$t_{\text{conctact}} = \left(\frac{312.5 N D_{\text{i}}}{L(T^2 H)^{1/3}}\right)^{-1}, \tag{129}$$

where N represents the agitation speed, D_{i} is the impeller diameter, whereas T and H are, respectively, the vessel diameter and vessel high.

According to Eq. (5), the most employed coalescence efficiency models for turbulent systems were derived from combination of the different expressions for drainage and contact times described so far. These were derived considering the compression force due to turbulent fluctuations, therefore this collision efficiency should be used with turbulent fluctuation collision mechanism. In fact, Chesters (1991) derived also the expressions for coalescence efficiency for viscous collision that can be used for velocity gradient or eddy capture collision mechanisms. However, Prince and Blanch (1990) used an identical efficiency for all collision mechanisms, whereas Wang et al. (2005a,b) stated that models for coalescence efficiency of different collision mechanisms might have different forms. In conclusion, if in a system different collision mechanisms are detected, it might be reasonable to estimate a different coalescence kernel for each collision mechanism, and in first approximation an overall coalescence kernel as sum of these can be computed (Liao et al., 2015).

6.2.4 Other Models for Contaminated Systems

The effect that contaminants have on the fluid particle interface can be accounted for with the previous models for partially mobile or immobile interfaces. However, other effects are not easy to be included in the model. For example, Podgórska and Marchisio (2016) in order to consider the repulsive electrostatic forces due to adsorption of ionic surfactant molecules on the interface, modified Eq. (101) in:

$$-\frac{\mathrm{d}g}{\mathrm{d}t}=\frac{2(F-F_{\mathrm{Rr}})}{3\pi\mu_{\mathrm{c}}r_{\mathrm{eq}}}g, \tag{130}$$

where F represents the turbulent compression force expressed by Eq. (105), while F_{Rr} is the repulsive force between spherical particles described as:

$$F_{\mathrm{Rr}}=\frac{64\pi R_{\mathrm{g}}k_{\mathrm{B}}\mathrm{T}n_{\infty}\gamma_0^2}{\kappa}\exp(-\kappa g), \tag{131}$$

with k_{B} Boltzmann constant, $n_{\infty}=N_{\mathrm{A}}M$, $1/\kappa$ is the Debye length and

$$\gamma_0=\frac{\exp(ze\Psi_0/2k_{\mathrm{B}}\mathrm{T}-1)}{\exp(ze\Psi_0/2k_{\mathrm{B}}\mathrm{T}-1)} \tag{132}$$

where Ψ_0 is the electrostatic surface potential. Integration of Eq. (130) from the initial film thickness g_i, to the critical g_f and neglecting the effect of intermittency, results in the following expression (Podgórska and Marchisio, 2016):

$$t_{\mathrm{drainage}}=\frac{3\pi\mu_{\mathrm{c}}r_{\mathrm{eq}}^2\ln(g_i/g_f)}{2[\pi r_{\mathrm{eq}}^2\rho_{\mathrm{c}}\langle\epsilon\rangle^{2/3}\xi_{1,2}-A_1r_{\mathrm{eq}}M\kappa^{-1}\exp(-\kappa g_f)]}, \tag{133}$$

where $A_1=64000\pi k_{\mathrm{B}}\mathrm{T}N_{\mathrm{A}}\gamma_0^2$. This relation was obtained for non-deformable (very small) spherical particles. In order to extend this approach for large particles, Podgórska and Marchisio (2016) considered the electrostatic repulsion force also in Eq. (108) for deformable particles with immobile interfaces:

$$-\frac{\mathrm{d}g}{\mathrm{d}t}=\frac{2g^3(F-F_{\mathrm{Rp}})}{3\pi\mu_{\mathrm{c}}R_a^4}, \tag{134}$$

where in this case F was evaluated with Eq. (109) and F_{Rp} represents the repulsion force between two planar double layers and is given by:

$$F_{\mathrm{Rp}}=64000\pi R_a^2k_{\mathrm{B}}TN_{\mathrm{A}}\gamma_0^2M\exp(-\kappa g). \tag{135}$$

Therefore, by integrating Eq (134) from g_i to g_f and by using the method of steepest descent following expression was obtained:

$$t_{\text{drainage}} = \frac{3\pi^2 \mu_c \rho_c r_{\text{eq}}^4 \epsilon^{2/3} \xi_{1,2}^{2/3} \left(g_f^{-2} - g_i^{-2}\right)\left(\xi_{1,2}/L\right)^{0.026}}{8\pi^2 \sigma\left(2\sigma - A_2 r_{\text{eq}} M \exp\left(-\kappa g_f\right)\right)}, \tag{136}$$

with $A_2 = A_1/\pi$.

Another different approach was adopted by Maindarkar et al. (2013) to take into account the emulsion stabilization effect due to nonionic surfactants. They modified the coalescence kernel of Coulaloglou and Tavlarides (1977), introducing a description of time-varying surfactant adsorption and its associated effects on particle surface coverage, interfacial tension, and particle coalescence. The interfacial tension increases as free surfactant was depleted to stabilize the particles surface (Maindarkar et al., 2013):

$$\sigma(t) = \sigma_0 - R_g T \Gamma_\infty \ln\left(1 + \frac{C_{sur}(t)}{C_{sur,1/2}}\right), \tag{137}$$

where σ_0 is the interfacial tension of the system without surfactants, Γ_∞ is the maximum surface coverage, $C_{sur}(t)$ is the surfactant concentration at instant t, and $C_{sur,1/2}$ is the surfactants concentration corresponding to half of the maximum surface coverage. The time-dependent free surfactant concentration was calculated by a surfactant mass balance as following (Maindarkar et al., 2013):

$$\frac{\mathrm{d}C_{sur}}{\mathrm{d}t} = -\frac{6\alpha_{\mathrm{d}}}{(1-\alpha_{\mathrm{d}})} \frac{\mathrm{d}\left(\frac{\Gamma}{\xi_{32}}\right)}{\mathrm{d}t}, \tag{138}$$

where Γ is the surface coverage of surfactant. Assuming a fast adsorption, the surface coverage can be estimated through a Langmuir isothermal:

$$\Gamma = \Gamma_\infty \left(\frac{C_{sur}(t)}{C_{sur}(t) + C_{sur,1/2}}\right). \tag{139}$$

Hence, the Coulaloglou and Tavlarides (1977) coalescence kernel was modified by using Eq. (137) for σ and multiplying the kernel expression with a prefactor as follower (Maindarkar et al., 2013):

$$a(\xi_1, \xi_2) = \left(1 - \frac{\Gamma}{\Gamma_\infty}\right)^2 a_{\mathrm{CT}}(\xi_1, \xi_2), \tag{140}$$

where a_{CT} represents the standard form of the Coulaloglou and Tavlarides (1977) coalescence kernel. Eq. (140) shows as the coalescence rate decrease with the adsorption of surfactants molecules that could inhibit the coalescence due to their geometric conformation.

6.3 Comparison Among Different Models for Coalescence Kernels

In this section the different models for coalescence will be compared, limiting the comparison to the estimation of the coalescence efficiency obtained through the film drainage model. Calculations refer to toluene–water liquid–liquid dispersion, in which it was supposed that the two colliding particles have the same size. A single value of the turbulent kinetic energy dissipation rate ($\epsilon = 0.1\ \mathrm{m}^2\ \mathrm{s}^{-3}$) was employed. Fig. 7 shows the trends of three different coalescence kernels derived by Chesters (1991), respectively, for particles deformable with immobile interfaces, deformable with partially mobile interfaces and deformable with fully mobile interfaces. The mobility of the interface can be correlated to the contamination degree of the system, in fact the interfaces of two colliding particles can be considered fully mobile in an uncontaminated system, whereas immobile in a highly contaminated system. That being said, there is not a quantitative way to correlate such models to a certain degree of contamination of the

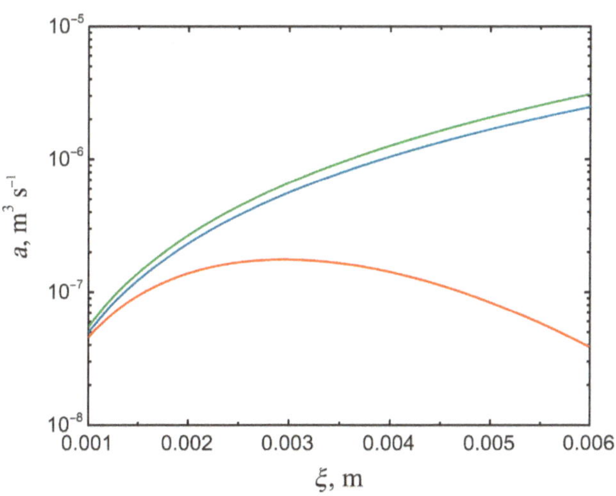

Fig. 7 Trends of coalescence kernels for: deformable particles with immobile interfaces (*red line*), deformable particles with partially mobile interfaces (*blue line*), deformable particles with fully mobile interface (*green line*) (Chesters, 1991).

system. The plot of the coalescence kernel for deformable particles with immobile interface shows a maximum value for a certain particles size, whereas the other two lines exhibit a growing trend in the range of particle size considered. It is interesting to note that the magnitude of the coalescence kernel increase, going from the deformable particles with immobile interfaces model to the deformable particles with fully mobile interfaces model. This is due to the time of drainage that is greater in the first case and lower in the second case. Therefore, for colliding particles with equal diameter the coalescence efficiency will be larger in the case of deformable particles with fully mobile interfaces, than for deformable particles with immobile interface and obviously it will have intermediate values for deformable particles with partially mobile interfaces. Fig. 8 shows the trends of different coalescence kernels for the same type of fluid particles. In general, it can be noted that in each subplot the trends and the orders of

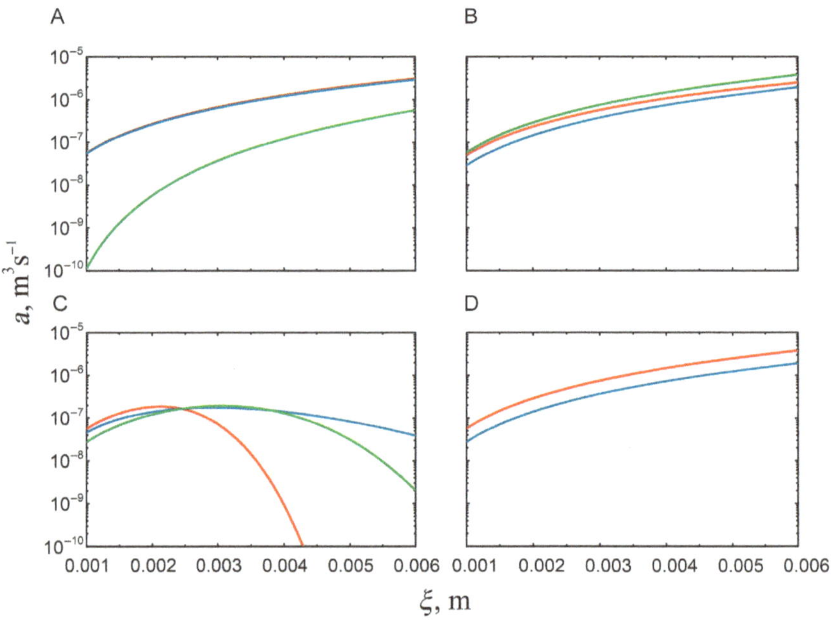

Fig. 8 Comparison of different coalescence kernel models. (A) Deformable particles with fully mobile interfaces: Chesters (1991) (*red line*), Prince and Blanch (1990) (*blue line*), and Lee et al. (1987a) (*green line*). (B) Deformable particles with partial mobile interfaces: Chesters (1991) (*red line*), Podgórska (2005) (*blue line*), and Lee et al. (1987a) (*green line*). (C) Deformable particles with immobile interfaces: Coulaloglou and Tavlarides (1977) (*red line*), Chesters (1991) (*blue line*), and Podgórska (2005) (*green line*). (D) Rigid particles: Chesters (1991) (*red line*) and Podgórska (2005) (*blue line*).

magnitude of the coalescence kernels are similar. The coalescence kernel models reported in Fig. 8A, B, and D show an increasing trend in the range of particle size analyzed. On the other hand, the trends of the coalescence kernel for deformable particles with immobile interface (Fig. 8C) show a maximum value in almost the same range of particles size. It is worth to point out that, also the coalescence kernels reported in Fig. 8A, B, and D should reach a maximum for a certain value of particle size, and this is not visible in the figure due to the limited range of particle size considered. In fact the coalescence efficiency estimated through the film drainage model normally tend to zero with the increase of particle size, therefore this model is not useful to describe coalescence of particles with relatively large size. For this reason, Lehr et al. (2002) developed the critical velocity approach model, and Sovová (1981) combined the drainage model of Coulaloglou and Tavlarides (1977) with the energy model given by Eq. (97). Fig. 9 regards the coalescence kernel for deformable and rigid particles with immobile interfaces that takes into account the presence of ions present in the continuous phase (Podgórska and Marchisio, 2016). In particular in the plot four curves are present: each of these was evaluated with a different concentration of a salt composed by monovalent ions. It is interesting to note how for colliding particles of equal size, the coalescence rate increases when the salt concentration grows. This happens because a higher concentration of salt decreases the Debye length and so the repulsive force between the electric

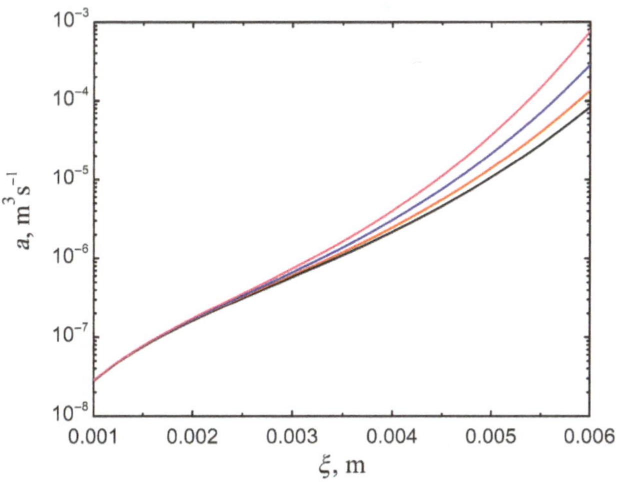

Fig. 9 Dependency of Podgórska and Marchisio (2016) coalescence kernel on different concentration of a salt composed by monovalent ions: 0.2 mM (*black line*), 0.3 mM (*red line*), 0.4 mM (*blue line*), and 0.5 mM (*magenta line*).

double layers of the colliding particles. In other words, the Van der Waals attractive forces exceed the electric repulsion forces from larger distances favoring the film rupture and coalescence.

Finally, it is worth highlighting that also in this case, as in the breakage kernel analysis (Section 4.4), the values of the empirical and semiempirical parameters employed in this analysis are those reported in the literature (Chesters, 1991; Coulaloglou and Tavlarides, 1977; Lee et al., 1987a; Podgórska, 2005; Podgórska and Marchisio, 2016; Prince and Blanch, 1990).

7. CONCLUSIONS

In this work, different kernel models were analyzed. In turbulent systems the prevalent breakage mechanism is caused by turbulent velocity fluctuations, whereas the others usually are neglected. To derive the breakage kernel for turbulent fluctuations different criteria were used, and consequently different expressions for breakage frequency were obtained. However, normally these expressions depend on the same variables, i.e., fluid particle size (ξ), flow characteristics (ϵ, k), physical–chemical proprieties of the system (σ, μ_d, μ_c, ρ_d, ρ_d), and empirical parameters. Another fundamental information needed are the number and the size of the particles resulting after breakage. For this reason, different approaches (empirical, statistical, and phenomenological) were employed to derive the daughter size distribution (DaSD). The phenomenological models are considered more physically consistent, and consequently they are the most used models for DaSD. In particular, among phenomenological models the M-shape daughter size distribution seems the most reasonable in case of bubbles and drops. The coalescence is intrinsically more complex than breakage, because it involves both the interaction between the fluid particles and continuous phase and also the interaction between fluid particles themselves. Normally, the coalescence kernel model is written in form of a product between a collision frequency and a coalescence efficiency. Also in this case, different collision mechanisms were considered to derive an appropriate model for each different type of collision. The most important collision mechanism in the turbulent system is caused by the turbulent velocity fluctuations. The presence of surfactants, salts, and other impurities in the systems influences the breakage and coalescence phenomena, and so the PSD. Therefore it is important to take into account the effects of the presence of contaminants in the coalescence and breakage kernel. The contaminants have greater effect on the coalescence than breakage phenomena, so in the breakage kernels only the interfacial tension is normally modified. In the coalescence kernel model the contaminations effects are take into account in the coalescence efficiency models. For instance, in the critical velocity model the value of critical approach velocity is decreased, whereas in the film drainage model is employed a different expression of drainage time depending by the contamination degree. In fact, the drainage time is normally derived considering the mobility of the interfaces of colliding particles. The interfaces mobility, in turn, depends of contamination degree of the systems, although there is not a mathematical relationship between these aspects. The presence of contaminants not only influences the mobility of particles interfaces, but also produce other effects that sometimes can depend of the specific type of contaminant. Therefore, some models are available to take into account specific effects of the contamination. To conclude, in last decades many models for the breakage and coalescence kernels were proposed, but they can be considered at the first stage of development. In fact, in the future to obtain consistent models, it will be necessary:

1. to replace the empirical or fitting parameters present in the models with universal constants by identify all dependencies;
2. to derive a breakage kernel that take into account all possible breakage mechanisms, and at same way to derive a coalescence kernel to consider all possible collision mechanisms for the fluid particles;
3. to consider the main effects of the contaminants in the kernels at any of contamination;
4. improve the experimental and numerical techniques that should help the modeler to derive and validate new and improved kernels.

REFERENCES

Alopaeus V, Koskinen J, Keskinen KI, Majander J: Simulation of the population balances for liquid-liquid systems in a nonideal stirred tank. Part 2—Parameter fitting and the use of the multiblock model for dense dispersions, *Chem Eng Sci* 57: 1815–1825, 2002.

Andersson R, Andersson B: On the breakup of fluid particles in turbulent flows, *AICHE J* 52:2020–2030, 2006.

Bak A, Podgórska W: Investigation of drop breakage and coalescence in the liquid-liquid system with nonionic surfactants tween 20 and tween 80, *Chem Eng Sci* 74:181–191, 2012.

Baldyga J, Bourne JR: Interpretation of turbulent mixing using fractals and multifractals, *Chem Eng Sci* 50:381–400, 1995.

Baldyga J, Bourne J: *Turbulent mixing and chemical reactions,* Chichester, UK, 1999, John Wiley & Sons.

Baldyga J, Podgórska W: Drop break-up in intermittent turbulence: maximum stable and transient sizes of drops, *Can J Chem Eng* 76:456–470, 1998.

Baldyga J, Podgorska W, Smit L: Turbulent dispersion of drops in intermittent turbulence. In *Proc. of 9th European conf. on mixing, mixing 97, multiphase systems, Nancy, France.* 1997, pp 247–254.

Bannari R, Kerdouss F, Selma B, Bannari A, Proulx P: Three-dimensional mathematical modeling of dispersed two-phase flow using class method of population balance in bubble columns, *Comput Chem Eng* 32:3224–3237, 2008.

Bentley BJ: A computer-controlled four-roll mill for investigations of particle and drop dynamics in two-dimensional linear shear flows, *J Fluid Mech* 167:219–240, 1986.

Carrica P, Clausse A: A mathematical description of the critical heat flux as a non-linear dynamic instability. In Gouesbet G, Berlemont A, editors: *Instabilities in multiphase flow*, New York, 1993, Plenum Press.

Casamatta G, Vogelpohl A: modelling of fluid dynamics and mass transfer in extraction columns, *Ger Chem Eng* 8:96–103, 1985.

Chatzi EG, Gavrielides AD, Kiparissides C: Generalized model for prediction of the steady-state drop size distributions in batch stirred vessels, *Ind Eng Chem Res* 28:1704–1711, 1989.

Chen XZ, Luo ZH, Yan WC, Lu YH, Ng IS: Three-dimensional CFD-PBM coupled model of the temperature fields in fluidized-bed polymerization reactors, *AICHE J* 57:3351–3366, 2011.

Chesters AK: The applicability of dynamic-similarity criteria to isothermal, liquid-gas, two-phase flows without mass transfer, *Int J Multiphase Flow* 2:191–212, 1975.

Chesters AK: Modelling of coalescence processes in fluid-liquid dispersions. a review of current understanding, *Chem Eng Res Des* 69:259–270, 1991.

Collins SB, Knudsen J: Drop-size distributions produced by turbulen pipe flow of immiscible liquids, *AICHE J* 16(6):1072–1080, 1970.

Coulaloglou C, Tavlarides L: Description of interaction processes in agitated liquid-liquid dispersions, *Chem Eng Sci* 32:1289–1297, 1977.

Craig VSJ, Ninham BW, Pashley RM: Effect of electrolytes on bubble coalescence, *Nature* 364:317–319, 1993.

Davis RH, Schonberg JA, Rallison JM: The lubrication force between two viscous drops, *Phys Fluids A* 1:77–81, 1989.

De Bruijn RA: Tipstreaming of drops in simple shear flows, *Chem Eng Sci* 48:277–284, 1993.

Doubliez L: The drainage and rupture of a non-foaming liquid film formed upon bubble impact with a free surface, *Int J Multiphase Flow* 17:783–803, 1991.

Duineveld PC, 1994. Bouncing and coalescence of two bubbles in water. Ph.D. thesis, University of Twente, The Netherlands.

Eastwood C, Cartellier A, Lasheras JC: The break-up time of a droplet in a fully developed turbulent flow. In *Advances in turbulence VII, Eight European turbulence conference,* 2000, Barcelona, pp 573–576.

Elemans PHM, Bos HL, Janssen JMH, Meijer HEH: Transient phenomena in dispersive mixing, *Chem Eng Sci* 48:267–276, 1993.

Friedlander SK: *Smoke, dust and haze,* New York, 1977, Wiley.

Fu X, Ishii M: Two-group interfacial area transport in vertical air–water flow: I. Mechanistic model, *Nucl Eng Des* 219:143–168, 2003a.

Fu XY, Ishii M: Two-group interfacial area transport in vertical air-water flow—II. Model evaluation, *Nucl Eng Des* 219:169–190, 2003b.

Gäbler A, Wegener M, Paschedag AR, Kraume M: The effect of pH on experimental and simulation results of transient drop size distributions in stirred liquid-liquid dispersions, *Chem Eng Sci* 61:3018–3024, 2006.

Ghasempour F, Andersson R, Andersson B: Multidimensional turbulence spectra–statistical analysis of turbulent vortices, *Appl Math Model* 38:4226–4237, 2014a.

Ghasempour F, Andersson R, Andersson B, Bergstrom DJ: Number density of turbulent vortices in the entire energy spectrum, *AICHE J* 60:3989–3995, 2014b.

Grace HP: Dispersion phenomena in high viscosity immiscible fluid systems and application of static mixers as dispersion devices in such systems, *Chem Eng Commun* 14:225–277, 1982.

Han L, Gong S, Ding Y, Fu J, Gao N, Luo H: Consideration of low viscous droplet breakage in the framework of the wide energy spectrum and the multiple fragments, *AICHE J* 61:2147–2168, 2015.
Henry CL, Dalton CN, Scruton L, Craig VSJ: Ion-specific coalescence of bubbles in mixed electrolyte solutions, *J Phys Chem C* 111:1015–1023, 2007.
Hesketh RP, Fraser Russell TW, Etchells AW: Bubble size in horizontal pipelines, *AICHE J* 33:663–667, 1987. Cited By 77.
Hesketh RP, Etchells AW, Russell TWF: Experimental observations of bubble breakage in turbulent flow, *Ind Eng Chem Res* 30:835–841, 1991.
Hibiki T, Ishii M: One-group interfacial area transport of bubbly flows in vertical round tubes, *Int J Heat Mass Transf* 43:2711–2726, 2000a.
Hibiki T, Ishii M: Two-group interfacial area transport equations at bubbly-to-slug flow transition, *Nucl Eng Des* 202:39–76, 2000b.
Hibiki T, Takamasa T, Ishii M: Interfacial area transport of bubbly flow in a small diameter pipe, *J Nucl Sci Technol* 38:614–620, 2001.
Howarth WJ: Measurement of coalescence frequency in an agitated tank, *AICHE J* 13:1007–1013, 1967.
Hsia MA, Tavlarides LL: Simulation analysis of drop breakage, coalescence and micromixing in liquid-liquid stirred tanks, *Chem Eng J* 26:189–199, 1983.
Kalkach-Navarro S, Lahey RT Jr., Drew DA: Analysis of the bubbly/slug flow regime transition, *Nucl Eng Des* 151:15–39, 1994.
Kamp AM, Chesters AK, Colin C, Fabre J: Bubble coalescence in turbulent flows: a mechanistic model for turbulence-induced coalescene applied to microgravity bubbly pipe flow, *Int J Multiphase Flow* 27:1363–1396, 2001.
Kennard EH: *Kinetic theory of gases, with an introduction to statistical mechanics,* New York, 1938, McGraw-Hill.
Kentish SE, Stevens GW, Pratt HRC: Estimation of coalescence and breakage rate constants within a Kühni column, *Ind Eng Chem Res* 37:1099–1106, 1998.
Kocamustafaogullari G, Ishii M: Foundation of the interfacial area transport equation and its closure relations, *Int J Heat Mass Transf* 38:481–493, 1995.
Kolmogorov AN: The local structure of turbulence in incompressible viscous fluid for very large reynolds numbers. In *Dokl Akad Nauk SSSR* 30:1941, pp 301–305. 4.
Konno M, Muko T, Saito S: Coalescence of dispersed drops in an agitated tank, *J Chem Eng Jpn* 21:335–338, 1988.
Koshy A, Das TR, Kumar R: Effect of surfactants on drop breakage in turbulent liquid dispersions, *Chem Eng Sci* 43:649–654, 1988.
Kuboi R, Komasawa I, Otake T: Collision and coalescence of dispersed drops in turbulent liquid flow, *J Chem Eng Jpn* 5:423–424, 1972.
Kumar MK, Mitra T, Ghosh P: Adsorption of ionic surfactants at liquid-liquid interfaces in the presence of salt: application in binary coalescence of drops, *Ind Eng Chem Res* 45:7135–7143, 2006.
Laakkonen M, Alopaeus V, Aittamaa J: Validation of bubble breakage, coalescence and mass transfer models for gas-liquid dispersion in agitated vessel, *Chem Eng Sci* 61:218–228, 2006. Cited By 71.
Laakkonen M, Moilanen P, Alopaeus V, Aittamaa J: Modelling local bubble size distributions in agitated vessels, *Chem Eng Sci* 62:721–740, 2007.
Lasheras JC, Eastwood C, Montañés C, Montaes JL: A review of statistical models for the break-up an immiscible fluid immersed into a fully developed turbulent flow, *Int J Multiphase Flow* 28:247–278, 2002.
Lee CH, Erickson LE, Glasgow LA: Bubble breakup and coalescence in turbulent gas-liquid dispersions, *Chem Eng Commun* 59:65–84, 1987a.

Lee CH, Erickson LE, Glasgow LA: Dynamics of bubble size distribution in turbulent gas-liquid dispersions, *Chem Eng Commun* 61:181–195, 1987b.
Lehr F, Mewes D: A transport equation for the interfacial area density in two-phase flow. In *Second European congress of chemical engineering*. France, 1999, Montpellier.
Lehr F, Millies M, Mewes D: Bubble-size distributions and flow fields in bubble columns, *AICHE J* 48:2426–2443, 2002. Cited By 194.
Levich VG: *Physicochemical hydrodynamics,* Englewood Cliffs, NJ, 1962, Prentice Hall.
Liao Y, Lucas D: A literature review of theoretical models for drop and bubble breakup in turbulent dispersions, *Chem Eng Sci* 64:3389–3406, 2009.
Liao Y, Lucas D: A literature review on mechanisms and models for the coalescence process of fluid particles, *Chem Eng Sci* 65:2851–2864, 2010. Cited By 162.
Liao Y, Lucas D, Krepper E, Schmidtke M: Development of a generalized coalescence and breakup closure for the inhomogeneous MUSIG model, *Nucl Eng Des* 241:1024–1033, 2011.
Liao Y, Rzehak R, Lucas D, Krepper E: Baseline closure model for dispersed bubbly flow: bubble coalescence and breakup, *Chem Eng Sci* 122:336–349, 2015.
Lin Y, Lee K, Matsoukas T: Solution of the population balance equation using constant-number Monte Carlo, *Chem Eng Sci* 57:2241–2252, 2002.
Luo H, 1993. Coalescence, break-up and liquid circulation in bubble column reactors. Ph.D. thesis, The Norwegian Institute of Technology, Trondheim, Norway.
Luo H, Svendsen H: Theoretical model for drop and bubble breakup in turbulent dispersions, *AICHE J* 42:1225–1233, 1996.
Maaß S, Metz F, Rehm T, Kraume M: Prediction of drop sizes for liquid-liquid systems in stirred slim reactors—Part I: Single stage impellers, *Chem Eng J* 162:792–801, 2010.
Maindarkar SN, Bongers P, Henson MA: Predicting the effects of surfactant coverage on drop size distributions of homogenized emulsions, *Chem Eng Sci* 89:102–114, 2013.
Marchisio D, Fox R: Solution of population balance equations using the direct quadrature method of moments, *J Aerosol Sci* 36:43–73, 2005.
Marchisio DL, Fox RO: *Computational models for polydisperse particulate and multiphase systems,* Cambridge, UK, 2013, Cambridge University Press.
Marchisio D, Pikturna J, Fox R, Vigil R, Barresi A: Quadrature method of moments for population-balance equations, *AICHE J* 49:1266–1276, 2003.
Martinez-Bazan C, Montañés JL, Lasheras JC: On the breakup of an air bubble injected into a fully developed turbulent flow. Part 1. Breakup frequency, *J Fluid Mech* 401:157–182, 1999a.
Martinez-Bazan C, Montañés JL, Lasheras JC: On the breakup of an air bubble injected into a fully developed turbulent flow. Part 2. Size pdf of the resulting daughter bubbles, *J Fluid Mech* 401:183–207, 1999b.
Martinez-Bazan C, Rodríguez-Rodríguez J, Deane GB, Montaes JL, Lasheras JC: Considerations on bubble fragmentation models, *J Fluid Mech* 661:159–177, 2010.
Narsimhan G: Model for drop coalescence in a locally isotropic turbulent flow field, *J Colloid Interface Sci* 272:197–209, 2004. Cited By 27.
Narsimhan G, Gupta JP, Ramkrishna D: A model for transitional breakage probability of droplets in agitated lean liquid-liquid dispersions, *Chem Eng Sci* 34:257–265, 1979.
Nguyen PT, Nguyen AV: Drainage, rupture, and lifetime of deionized water films: effect of dissolved gases, *Langmuir* 26:3356–3363, 2010.
Oolman TO, Blanch HW: Bubble coalescence in stagnant liquids, *Chem Eng Commun* 43:237–261, 1986.
Park JY, Blair LM: The effect of coalescence on drop size distribution in an agitated liquid-liquid dispersion, *Chem Eng Sci* 30:1057–1064, 1975.
Podgórska W: Scale-up effects in coalescing dispersions–comparison of liquid-liquid systems differing in interface mobility, *Chem Eng Sci* 60:2115–2125, 2005.

Podgórska W, Baldyga J: Scale-up effects on the drop size distribution of liquid–liquid dispersion in agitated vessels, *Chem Eng Sci* 56:741–746, 2001.
Podgórska W, Marchisio DL: Modeling of turbulent drop coalescence in the presence of electrostatic forces, *Chem Eng Res Des* 108:30–41, 2016.
Pope SB: *Turbulent flows,* Cambridge, UK, 2000, Cambridge University Press.
Prince M, Blanch H: Bubble coalescence and break-up in air-sparged bubble-columns, *AICHE J* 36:1485–1499, 1990.
Prince MJ, Blanch HW: Bubble coalescence and break-up in air-sparged bubble columns, *AICHE J* 36:1485–1499, 1990.
Randolph AD: Effect of crystal breakage on crystal size distribution in a mixed suspension crystallizer, *Ind Eng Chem Fundam* 8:58–63, 1969.
Renardy YY, Cristini V: Effect of inertia on drop breakup under shear, *Phys Fluids* 13:7–13, 2001a.
Renardy YY, Cristini V: Scalings for fragments produced from drop breakup in shear flow with inertia, *Phys Fluids* 13:2161–2164, 2001b.
Renardy Y, Cristini V, Li J: Drop fragment distributions under shear with inertia, *Int J Multiphase Flow* 28:1125–1147, 2002.
Ross SL, Verhoff FH, Curl RL: Droplet breakage and coalescence processes in an agitated dispersion. 2. Measurement and interpretation of mixing experiments, *Ind Eng Chem Fundam* 17:101–108, 1978.
Sagert NH, Quinn MJ: The coalescence of H_2S and CO_2 bubbles in water, *Can J Chem Eng* 54:392–398, 1976.
Sajjadi B, Raman AAA, Shah RSSRE, Ibrahim S: Review on applicable breakup/coalescence models in turbulent liquid-liquid flows, *Rev Chem Eng* 29:131–158, 2013.
Schwartzberg HG, Treybal RE: Fluid and particle motion in turbulent stirred tanks: particle motion, *Ind Eng Chem Fundam* 7:6–12, 1968.
Simon M, 2004. Koaleszenz von tropfen und tropfenschwärmen. Ph.D. thesis, die Teschinschen Universtität Kaiserslautern.
Solsvik J, Jakobsen HA: Development of fluid particle breakup and coalescence closure models for the complete energy spectrum of isotropic turbulence, *Ind Eng Chem Res* 55:1449–1460, 2016.
Solsvik J, Tangen S, Jakobsen HA: On the constitutive equations for fluid particle breakage, *Rev Chem Eng* 29:241–356, 2013.
Solsvik J, Skjervold VT, Jakobsen HA: A bubble breakage model for finite Reynolds number flows, *J Dispers Sci Technol* 38:973–978, 2017.
Sovová H: Breakage and coalescence of drops in a batch stirred vessel-II comparison of model and experiments, *Chem Eng Sci* 36:1567–1573, 1981.
Su J, Gu Z, Li Y, Feng S, Xu XY: An adaptive direct quadrature method of moment for population balance equations, *AICHE J* 54:2872–2887, 2008.
Sun X, Kim S, Ishii M, Beus SG: Modeling of bubble coalescence and disintegration in confined upward two-phase flow, *Nucl Eng Des* 230:3–26, 2004.
Tavlarides LL, Stamatoudis M: The analysis of interphase reactions and mass transfer in liquid-liquid dispersions, *Adv Chem Eng* 11:199–273, 1981.
Tsouris C, Tavlarides LL: Breakage and coalescence models for drops in turbulent dispersions, *AICHE J* 40:395–406, 1994.
Valentas KJ, Amundson NR: Breakage and coalescence in dispersed phase systems, *Ind Eng Chem Fundam* 5:533–542, 1966.
Verhoff FH, Ross SL, Curl RL: Breakage and coalescence processes in an agitated dispersion. Experimental system and data reduction, *Ind Eng Chem Fundam* 16(3): 371–377, 1977.
Walter JF, Blanch HW: Bubble break-up in gas-liquid bioreactors: break-up in turbulent flows, *Chem Eng J* 32:B7–B17, 1986.

Wang T, Wang J: Numerical simulations of gas-liquid mass transfer in bubble columns with a CFD-PBM coupled model, *Chem Eng Sci* 62:7107–7118, 2007.
Wang T, Wang J, Jin Y: A novel theoretical breakup kernel function for bubbles/droplets in a turbulent flow, *Chem Eng Sci* 58:4629–4637, 2003.
Wang T, Wang J, Jin Y: Population balance model for gas–liquid flows: influence of bubble coalescence and breakup models, *Ind Eng Chem Res* 44:7540–7549, 2005a. Cited By 71.
Wang T, Wang J, Jin Y: Theoretical prediction of flow regime transition in bubble columns by the population balance model, *Chem Eng Sci* 60:6199–6209, 2005b.
Wieringa JA, Van Dieren F, Janssen JJM, Agterof WGM: Droplet breakup mechanisms during emulsification in colloid mills at high dispersed phase volume fraction, *Chem Eng Res Des* 74:554–562, 1996.
Wright H, Ramkrishna D: Factors affecting coalescence frequency of droplets in a stirred liquid–liquid dispersion, *AICHE J* 40:767–776, 1994.
Wu Q, Kim S, Ishii M, Beus SG: One-group interfacial area transport in vertical bubbly flow, *Int J Heat Mass Transf* 41:1103–1112, 1998.
Yan WC, Li J, Luo ZH: A CFD-PBM coupled model with polymerization kinetics for multizone circulating polymerization reactors, *Powder Technol* 231:77–87, 2012.
Yao Y, Su JW, Luo ZH: CFD-PBM modeling polydisperse polymerization FBRs with simultaneous particle growth and aggregation: the effect of the method of moments, *Powder Technol* 272:142–152, 2015.
Yuan C, Fox RO: Conditional quadrature method of moments for kinetic equations, *J Comput Phys* 230:8216–8246, 2011.
Yuan C, Laurent F, Fox RO: An extended quadrature method of moments for population balance equations, *J Aerosol Sci* 51:1–23, 2012.
Zaccone A, Gäbler A, Maaß S, Marchisio D, Kraume M: Drop breakage in liquid-liquid stirred dispersions: Modelling of single drop breakage, *Chem Eng Sci* 62:6297–6307, 2007.
Zhao H, Ge W: A theoretical bubble breakup model for slurry beds or three-phase fluidized beds under high pressure, *Chem Eng Sci* 62:109–115, 2007.

INDEX

Note: Page numbers followed by "*f*" indicate figures, and "*t*" indicate tables.

A

Adaptive direct quadrature method of moments (ADQMOM), 127
Air–methane combustion, 57–58
Air–water system, 157–159
Anisotropic Gaussian (AG) distribution, 34–35

B

Bidisperse gas–particle flow, 4*f*
Boussinesq approximation
  fluid–particle velocity correlations, 119–121
  particle stress tensor, 117–119
Breakage frequency model, 154
Breakage kernel
  in contaminated systems, 156–157
  model
    comparison, 157–159, 157*f*
    multifractal model, 150–154
Breakage mechanism model, 155–156
Bubble size distribution (BSD), 126

C

CFD. *See* Computational fluid dynamics (CFD)
CHyQMOM, 16–17
CIT. *See* Cluster-induced turbulence (CIT)
Classes method (CM), 127
Cluster-induced turbulence (CIT), 33
  Lagrangian simulations of, 34*f*
Coalescence, 128, 134–135
  efficiency model, 168
    critical approach velocity model, 169
    energy model, 168–169
    film drainage model, 169–176, 170*f*
  kernels model, 179–182, 179–181*f*
Collision frequency, 8
Collision frequency model, 164
  collisions due to
    mechanisms, 167–168
    turbulent velocity fluctuations, 164–167
Composition moment transport equation, 9
Computational fluid dynamics (CFD), 2–3, 126
  macroscale models, 6*f*
Conditional quadrature methods of moments (CQMOM), 15–16, 127
Contaminated system
  breakage kernel in, 156–157
  models for, 177–179
Contamination effects, 156–157
CQMOM. *See* Conditional quadrature method of moments (CQMOM)

D

Daughter size distribution function (DaSD) model, 159
  phenomenological models, 160–163
    bell-shape, 160–161
    M-shape, 162–163
    U-shape, 161–162
  statistical models, 159–160
Deformable particles
  with fully mobile interfaces, 173–176
  with immobile interfaces, 172
  with partially mobile interfaces, 173
Delta-shocks, 15
Dilute-to-dense particle flows, gas-particle flow model for, 37–38
Direct quadrature method of moments (DQMOM), 127
Dispersed phase modeling, 113–114
DLVO theory, 135
DQMOM. *See* Direct quadrature method of moments (DQMOM)
Droplet size distribution (DSD), 126
Dryer and Glassman (DG) mechanism, 105–106, 105*t*

E

Eddy Break-Up (EBU) model, 104–105
Eddy capture mechanism, 132–133
Eddy Dissipation Concept (EDC), 104–105
Eddy Dissipation Model (EDM), 104–105

Energy model, 133–134, 168–169
Enskog kinetic theory of gases, 117–119
EQMOM. *See* Extended quadrature method of moments (EQMOM)
Erosive breakage, 131
Eulerian approach, 60–61
  dispersed phase, 113–114
  gas phase, 110–112
Eulerian moment
  methods, 33
  transport equation, 9
Eulerian simulations, 33–35, 36*f*
Extended quadrature method of moments (EQMOM), 14–15, 127

F

Film drainage model, 133–135, 134*f*, 169–176, 170*f*
Fine particles, 2–3
  in laminar flows, 31–32
  population balance equation for, 7, 30–33
  in turbulent flows, 32–33
Finite Reynolds number, particle breakage model for, 154–155
Finite-Stokes-number particles, 10
First-order upwind method, 17
Fluidized bed reactors, 55–57. *See also* Multiphase reactive flows
  air–methane combustion, 57–58
  gaseous combustion, 55–57
  3D numerical simulations, 72–75
Fluidized-bed simulation, 43*f*
Fluid particle
  breakage kernel model, 136–159
  breakup, phenomenology, 128–131
  coalescence
    kernel model, 163–182
    phenomenology, 131–135
  deformation, 130*f*
  flows, 6–8
  possible collision mechanisms for, 132*f*
  turbulent kinetic energy for, 137–139
  velocity correlations, 119–121
Fokker–Planck (FP) model, 24

G

Gaseous reactions, 55–57
Gas kinetic theory, 143–144
Gas–liquid system, 131, 135, 156
Gas–particle flow, 41–42
  governing equations for, 35–39
  kinetic equation for, 33–45
  model for dilute-to-dense particle flows, 37–38
  solution algorithm for, 39–42
Gas-phase chemical reactions, 104–105
Gas-phase turbulence modeling, 110–113
Gas–solid reactions, 110–112
Gaussian quadrature, 12
Gauss–Radau quadrature, 14–15, 32–33
Generalized population balance equation (GPBE), 10
  for polydisperse inertial particles, 8
Gibbs–Marangoni effect, 134–135
GPBE. *See* Generalized population balance equation (GPBE)

H

Homogeneous reactions, 87
Hydrodynamic model, 10
Hyperbolic quadrature method of moments (HyQMOM), 15–17
  hyperbolic spatial fluxes using, 20–21

I

Immobile interfaces
  deformable particles with, 172
  nondeformable particles with, 170–172
Integral collision, 114–115

K

KBFV. *See* Kinetic-based finite-volume (KBFV) methods
Kernal density function (KDF), 14–15
Kinetic-based finite-volume (KBFV) methods, 17–22
  hyperbolic spatial fluxes using HyQMOM, 20–21
  kinetic-based spatial fluxes, 19–20
  moment transport equations solutions, 18–19
  realizable time-stepping schemes, 21–22
Kinetic-based spatial fluxes, 19–20
Kinetic collision energy, 133–134
Kinetic energy, pseudo-turbulent, 4*f*
Kinetic equation (KE)
  collision-less, 11*f*
  for gas–particle flows, 33–45
  for monodisperse inertial particles, 8

Kinetic flux-splitting scheme, 39–40
Kinetic-theory modeling approach, 5, 6*f*
Kolmogorov theory, 127

L

Lagrangian simulations, 33–35, 36*f*
  of cluster-induced turbulence, 34*f*
Laminar flow, fine particles in, 31–32
Liquid–liquid dispersion, 138, 144–145

M

Mach equations, 110–112
Macroscale computational fluid dynamics
  models, 6*f*
  simulation, 23*f*
Macroscale model, 9
  tractable, 5–6
Maxwell–Boltzmann distribution, 142–143
Mean interphase enthalpy transfer, 115–117
Mesoscale model, 5–8, 6*f*
Methane–air premixed combustion, 61–62
Microscale direct-numerical simulation (DNS), 6*f*
Mobile interfaces
  deformable particles with fully, 173–176
  deformable particles with partially, 173
Moment methods, 9–11, 11*f*
Moment transport equation, 10, 18–19, 38–39
Monodisperse inertial particles, kinetic equation for, 8
Monte Carlo method (MCM), 127
Multifractal theory, 150–151
Multiphase reactive flows
  air–methane combustion, 57–58
  applications, 55–57
  dispersed phase modeling, 113–114
  Eulerian approach, 60–61
  fluidized bed reactors, 55–57
  fluid–particle velocity correlations, 119–121
  gas-phase modeling, 110–113
  hydrodynamic interactions, 55–57
  integral collision, 114–115
  local phase temperature determination, 64–65
  mean interphase transfer modeling, 115–117
  methane–air premixed combustion, 61–62
  natural gas combustion
    conductive heat exchanges, 79–81
    grid refinement, effect of, 81–86
    macroscopic and mesoscopic computation, 87–104
    preliminary results, 77–79
    subgrid turbulent mixing effect, 104–108, 107*f*
  NEPTUNE_CFD
    bi-disperse test case, 66–69, 69*f*
    characteristics, 65–66
    evolution, 71–72, 72*f*
    gas properties, reactor inlet, 75, 76*t*
    gas–solid computations, 71–72, 72*f*
    mesh characteristics, 73*t*
    mono-disperse test case, 66–69, 68*f*
    numerical configuration, 72–73, 73*f*
    parallel performances, 69–71, 71*f*
    particle properties, 74–75, 75*t*
    powder/gas properties, 70*t*
    simulation test cases, 74, 74*t*
    3D granular uniform shear flow, 66, 67*f*
    3D unsteady dense fluidized bed, 69–71, 70*f*
  particle stress tensor modeling, 117–119
  thermal radiation model, 63–64, 64*t*
Multiscale model
  approach, 5–9, 6*f*
  for polydisperse fluid–particle flows, 3–5, 4*f*

N

Navier–Stokes equation, 5–6, 8–9
Nondeformable particle, with immobile interfaces, 170–172
Number density function (NDF), 7–8, 17*f*, 126

O

1-D moment advection equation, 19
Operator-splitting algorithm, 40–41

P

Particle
  breakage model for finite Reynolds number, 154–155
  stress tensor modeling, 117–119
  turbulent velocity fluctuation around, 139–141
Particle-laden flows, application to, 42–45

Particle–particle collisions, 5
Particle–particle interaction, 5
Particle size distribution (PSD), 5, 126
Particle volume fraction, probability density function, 36*f*
PBE. *See* Population balance equation (PBE)
PBM. *See* Population balance model (PBM)
PDF. *See* Probability density function (PDF)
Phenomenology
  fluid particle breakup, 128–131
  fluid particle coalescence, 131–135
Polydisperse fluid–particle flows
  multiscale models for, 3–5, 4*f*
  numerical simulation of, 3–4
Polydisperse inertial particles, population balance equation for, 8
Population balance equation (PBE), 126
  for fine particles, 7, 30–33
Population balance model (PBM), 127–129
Probability density distribution function, 140
Probability density function (PDF), 5–6, 22
  composition PDF transport equation, 6
  kinetic equation, 8
  particle volume fraction, 36*f*
  population balance equation, 7
  transport equation, 22–23
PSD. *See* Particle size distribution (PSD)
Pseudo-turbulent kinetic energy, 4*f*

Q

Quadrature-based moment method (QBMM), 2–3, 11*f*, 12–17, 30, 127
  extended quadrature method of moments, 14–15
  hyperbolic quadrature method of moments, 15–17
  implemented in OpenFOAM, 45*f*
  quadrature method of moments, 13–14
Quadrature method of moments (QMOM), 127

R

Rayleigh–Taylor instability, 130–131
Robust inversion algorithm, 12

S

Shearing-off process, 131, 156
Stefan–Boltzmann constant, 63–64
Stokes drag, 9
Subgrid turbulent mixing, 107*f*
Surfactants, 131
  influence of, on fluid particle breakage, 127–128

T

Thermal radiation model, 63–64, 64*t*
3D CFD. *See* Multiphase reactive flows
3-D wall-bounded vertical channel, 42
Toluene–water system, 157–159
Tractable macroscale model, 5–6
Turbulent damping factor, 144–145
Turbulent diffusivity coefficient, 117–119
Turbulent flow
  collisions, 132
  fine particles in, 32–33
Turbulent fluctuation
  breakage models, 136–155
  mechanism, 129, 136
Turbulent inertial stress, of hitting eddy, 147–150
Turbulent kinetic energy, for fluid particle, 137–139
Turbulent kinetic of bombarding eddy, 141–147
Turbulent mixing, combustion, 104–108, 107*f*
Turbulent reacting flow, 22–30, 23*f*
  algorithm for, 28–30
  application to
    multivariate case with chemical reactions, 26–28
    univariate case without reactions, 24–25, 25*f*
Turbulent-Reynolds stress tensor, 110–112
Turbulent velocity fluctuation, around particle, 139–141
Turbulent viscosity, 110–112
2-D velocity phase space, 16

V

Velocity
  gradient mechanism, 132–133
  model, critical approach, 133–134
Viscous shear forces, 130

FSC
www.fsc.org
MIX
Papier | Fördert gute Waldnutzung
FSC® C083411

Kolibri 360 GmbH
Ferdinand-Jühlke-Straße 7
99095 Erfurt, Deutschland
produktsicherheit@kolibri360.de